# LION
# HEARTED

## THE LIFE AND DEATH OF CECIL &
## THE FUTURE OF AFRICA'S ICONIC CATS

ANDREW LOVERIDGE

NEW YORK

Regan Arts.

**Regan Arts.**

New York, NY

First Regan Arts hardcover edition, April 2018.

Library of Congress Control Number: 2018932425

ISBN 978-1-68245-120-5

Jacket and interior design by Lynne Ciccaglione

Image credits, which constitute an extension of this
copyright page, appear on page 280.

Printed in the United States of America

10 9 8 7 6 5 4 3 2 1

Previous page: Lioness in the Dete Valley.

Linkwasha, Hwange National Park.

This book is dedicated to my father, John Loveridge, who taught me to care about Africa's wild places, and to Joanne, Michael, and Ciara, without whose love, support, and indulgence this book would never have been written.

Cecil, 2011.

# CONTENTS

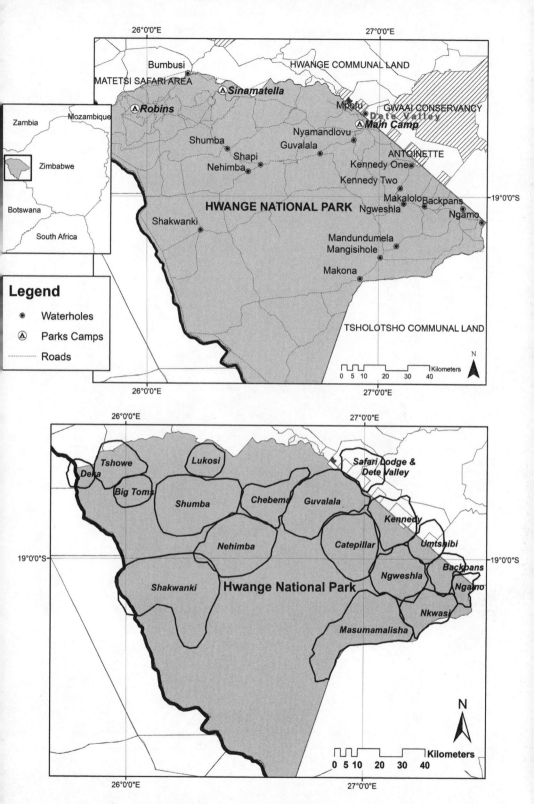

Locations of the main lion prides studied in Hwange National Park.

Cecil charging the research vehicle
seconds after he was first darted.

# PROLOGUE

## A LION CALLED CECIL

A stone's throw away, the reclining male lion moved restlessly as I edged the four-wheel-drive Land Cruiser forward across the uneven ground. My wife, Joanne, and I had been slowly maneuvering closer to him for the last half hour, stopping the vehicle whenever he became uneasy. Our cautious advance was gradually allowing the big cat to become accustomed to our presence. The truck's tires crunched over the dry grass tufts. The pupils of the lion's yellowish eyes widened as the potentially threatening vehicle approached. His black-tipped tail twitched a silent warning. Twenty meters away from him, I cut the engine once again and we sat motionless, giving him time to relax, making no movement or noise. As a newcomer to the area, he was not habituated to vehicles or people and could behave aggressively, or worse, run away before we could catch and tag him with a radio collar. Once fitted, the collar, with a radio transmitter and sophisticated global-positioning system, would log his location every hour and allow us to study his ranging movements and behavior.

Joanne held the loaded dart rifle, ready to pass it to me as soon as I saw an opportunity to shoot a tranquilizer dart into the wary lion. To be sure of a clean shot, we needed to be close to the lion and ensure there were no grass stalks that could deflect the light dart. After a few minutes intently scrutinizing the intruding vehicle, the lion relaxed again, shifting position to scan the open plain and the bush under which his brother was mating with one of the pride females. The change in position gave us a chance for a shot. Through the vehicle's open window, I slowly raised the rifle to my shoulder, bringing the red dot of its optical scope to bear on the lion's muscular, tawny body. A gentle squeeze of the trigger and a *phut* of compressed air sent the dart punching into the cat's shoulder. He jumped in surprise and gave a guttural snarl of annoyance. He glared furiously at the vehicle and, with a throaty warning growl, rushed forward a few meters in a tail-lashing mock charge. Teeth bared, he stopped, attention fixed on us, waiting for a cue to unleash a final assault on the vehicle. We froze. Any careless movement could precipitate a full-on attack. However, with nothing to provide a focus for his anger, his attention wandered. Honor satisfied, he stalked off stiff-legged to a nearby termite mound to nurse his wounded dignity. We sighed as the tension of the last half hour was released, the adrenaline of stalking and darting the lion ebbing from our bodies. With the dart in, we now had to wait for the drugs to take effect on the big cat. After five minutes he started to look groggy. After fifteen minutes his great head slumped onto his enormous paws. It was now safe to approach and handle him.

It was November 8, 2008, and the Lion Research Project had been running in Zimbabwe's Hwange National Park for nearly a decade. The project had been studying the population dynamics of lions and had by this point already collared more than 160 of them. This male, though, was unique—destined to become one of the most recognizable wild animals in the world. That day, in the early morning sunshine, deep in Hwange National Park on the Makalolo plain, dotted with gnarled *mitswiri* trees, we had no notion of this.

Seeing a lion this close up is an experience that even seasoned wildlife professionals don't often get. Despite having studied these animals for nearly twenty years, I still marvel at their sheer size. A male is close

2

to three meters long from nose to tail tip and stands 1.2 meters at the shoulder. A large male lion tips the scales at around 225 kilograms (495 pounds), the combined weight of two of the largest American football players. A male's size is augmented by a thick mane of dense fur on his head, chest, and shoulders—a signal of strength and virility that also serves as protection from raking claws in territorial fights with other males. Male lions are designed for war. They are the battle tanks of the cat world. Their adult lives are spent defending their territories and prides of females against other males. Lionesses are about 30 percent smaller, lithe, and muscular but no less formidable. A lioness has evolved as a huntress; her speed and raw strength ripple through every inch of her honey-colored body. When handling an immobilized lion, I often remind myself of the sheer size of these creatures by resting my hand against the underside of the soup plate–sized paw. A lion's paw is pretty much the same size as a man's hand, fingers splayed. Five lethally sharp claws, each three centimeters long, are safely sheathed within the paw. They are used as grappling hooks to pull down prey many times a lion's size, or as slicing, cutthroat razors in territorial battles. These animals are beautiful and dangerous and perfectly adapted to fulfill their role as the ultimate predator on the African plains.

Yet, despite their magnificence, they are under real threat. Their numbers have declined precipitously over the last decade. Across West Africa, where lions were historically widespread, there may be as few as 400 remaining. Lions are doing similarly badly across Central Africa, where insecurity and conflict have undermined the safety of their previous strongholds. It is only in eastern and southern Africa that lions are relatively safe. But even there many populations are declining, some of them frighteningly fast.

This is startling given that lions are one of the most recognizable and beloved mammal species in the world. What child in the Western world has not empathized with young Simba, the central character of the Disney animation *The Lion King*, one of the highest-grossing films of

all time? Who has not enjoyed the spectacle of Alex, the Central Park Zoo lion in *Madagascar*, cavorting incongruously with a hippopotamus, a giraffe, a zebra, and four psychotic penguins? The lion's image and reputation for strength and bravery have been appropriated by royalty, capital cities, sports teams, and multinational corporations. I'd hazard a guess that there are more images of lions in sculptures, carvings, gargoyles, and coats of arms in central London than there are living lions in the whole of West Africa. Surely, given their obvious cultural resonance, lions cannot be allowed to become yet another species balancing precariously on the brink of extinction? Or, worse, completely disappear. Yet this is what is happening. Lions will not disappear in the next few decades, or perhaps even in our lifetimes, but their slow, inexorable decline is a reality. And aside from a few concerned conservationists, humanity is doing little to prevent it. By the time the world wakes up to the enormity of what we are about to lose, it may already be too late.

We formed a corral of research vehicles around the sleeping cat to block the view of the other lions watching curiously from fifty meters away. Experienced hands fitted him with a radio collar, measured him, and took DNA and blood samples—all part of our scientific protocol. The necessary procedures complete, I injected the antidote to the tranquilizer and we backed the vehicles off to a safe distance to wait for our latest study animal to recover from his brief anesthesia. We needed a name for our newly collared male. The research project's study data are recorded against a scientifically stark identification code. Study lions are often also given nicknames, as it is more practical to use memorable names for everyday use. They are often named for a landscape feature in their territory, the circumstances of the first time they were seen or collared, or some unique physical attribute. Some are given dignified African names. A particularly regal lion was named Isikulu ("the Duke"). One sociable cat came to be known as Salibonani (which means "Hi There"). The research team dubbed one unfortunate pair

Cecil, December 2008.

of lionesses "Kick" and "Ass" in commemoration of their ignominious defeat by the herd of buffalo they were hunting. Many lions are named for their physical appearance or character. Mafuta ("the Fat One") was named for her over-full belly while Ulaka ("the Fierce One") earned his sobriquet by frequently charging tourist vehicles.

The lion we just collared had first been seen with his brother at the Mangisihole water hole, which translates roughly from the local language as "the Englishman's water hole." I have never discovered why the water hole was given this name. Perhaps it was named after an early English explorer or hunter traveling across the area in the nineteenth century, or maybe after one of the park's first game rangers. While we waited for the lion to wake, someone suggested, "Since he came from Mangisihole, we should give him a typical English name." After a short

Joanne and I fit a GPS collar to a lioness.

pause, someone said, "How about Cecil? That sounds really English." And so Cecil the Lion, member of the Mangisihole coalition, ID code MAGM1, began his career as a study animal on the Hwange Lion Research Project.[1]

When I first collared Cecil, he was between five and six years old, a male coming into his prime and challenging other males for territorial dominance. He was to live out his life in the East of Hwange, until the age of twelve, almost continuously monitored by the research project. From that first encounter, he became a favorite lion amongst our research team, photographic safari guides, and the tourists who visited the safari lodges his territory encompassed. Both he and his brother lost their initial wariness of people and could be easily approached by open-backed safari vehicles full of noisy tourists. One can only guess how often Cecil was photographed and videoed over the seven years he dominated the prides in the area. Perhaps thousands or tens of thousands of times, given that he lived in a prime photographic tourist area and was so well habituated to vehicles.

The routine capture and collaring of Cecil that still morning on the Makalolo Plains could not have contrasted more starkly to the international media frenzy caused by his death seven years later. The story of his demise at the hands of a rogue trophy hunter would dominate the front pages of the world's print media for weeks. *Time* magazine voted "Cecil the Lion" as number one among the "100 Most Influential Animals of 2016."[2] While this article was a tongue-in-cheek foil to the more sober "100 Most Influential People of 2016," the fact that Cecil was deemed the four-legged equivalent of Leonardo DiCaprio and Mark Zuckerberg tells us something about how sensational this story became. The small plastic dart that flew toward a then-anonymous lion was like a pebble thrown into a still pond. The ripples would, some years later, become a storm that would shake the foundations of the international trophy-hunting industry. It would also bring significant scrutiny of the way we conserve and value wild animals, awakening a global concern for Africa's iconic big cat.

A well-known Zimbabwean proverb holds that "Until the lion has its own storyteller, tales of the lion hunt will always glorify the hunter." This is the story of Cecil the Lion. It is also the story of my own experiences as a field biologist working in Africa and the realities of conserving large, dangerous animals on a continent whose ancient landscape is increasingly dominated by people. Cecil's story provides a lens through which to explore humanity's relationship with the natural world and what we can do to prevent this majestic species from disappearing.

# ONE

# THE VALLEY OF THE LIONS

People I meet frequently ask, with sometimes barely hidden incredulity, how I came to work with lions. The truth is that there is no one event that led me there. Perhaps my career had its beginnings in an African childhood. My dad, a zoologist, sparked and then nurtured my interest in animals—an interest later encouraged by several great conservationists: an Oxford professor, a professional hunter and guide, and a passionately dedicated National Parks[3] warden. Without the influence of these four men—each in his own way worried about the destruction of wildlife—the Hwange lion project would never have started. One particular experience, though, piqued an interest in lions and large carnivore conservation and put me on the pathway to Hwange National Park.

My introduction to the quintessential sound of wild Africa, the roar of a lion, came in 1992. I was on a camping trip with some university friends during summer vacation. We had stopped for the night at a roadside campsite on the way to Victoria Falls. It is folly to drive after dark in this part of the world; the roads are made hazardous by stray livestock, speeding trucks, and inebriated drivers in cars with no headlights. Our campsite, in a shady forest of Zambezi teak trees, was close to the head of the Dete Valley, in the wild, 1,000-square-kilometer Gwaai Conservation Area bordering Hwange National Park.

Lying in my sleeping bag, covered by just a net to keep away the malarial mosquitoes, I listened to the nighttime noises of Africa. The drowsy *prrrp, prrrp, prrrp* of the Scops owl. The bustle of small creatures in the undergrowth. The whoops of hunting hyenas, comfortably in the far distance.

Suddenly, my contented contemplation of the soft African night was shattered by a thundering bellow, the unmistakable call of a lion.

OOOOM . . . OOOOM . . . *ooom* . . . *ooom* . . . *ooom* . . . *om* . . . *om*.

The noise is raw and visceral, awakening the terror of the soft-skinned primeval hominid who lives in us all. It reminds us that, in the wild, the darkness belongs to the predator and that sharp tooth and curved claw reign until the dawn lights the sky. Most people's only exposure to a lion's roar is at the beginning of movies produced by Hollywood film studio Metro-Goldwyn-Mayer. Actually, a lion's roar sounds absolutely nothing like this. Embarrassingly for MGM, its leonine mascot's "roar" is reputed to be a recording of a snarling tiger.[4] A real lion's roar is an unforgettable sequence of reverberating, guttural grunts, generated deep in the big cat's chest and amplified by a complex series of folds and a flexible, partially ossified hyoid bone in its throat. Only cats in the Genus *Panthera*—lions, tigers, jaguars, leopards, and snow leopards—can make this sound, and of the five "roaring cat" species, the lion's roar is the most distinctive. The roar begins with a resonating bellow and tails off into a series of pugnacious, grunting *ooomfs*. The noise seems to vibrate through the ground, the echo bouncing through the bush and forest—a blast of noise that proclaims the ownership of all the land and the fearsome might of the owner.

The Somalis, from the Horn of Africa, have a proverb that says a brave man will be frightened by a lion three times. Once when he hears its voice; again when he sees its giant tracks; and, finally, when he glimpses the huge cat at close quarters for the first time. I can attest to the truth of the first statement. It is easy to feel insecure separated from a vocal lion by only a flimsy mosquito net. This lion sounded, to my inexperienced ears, perhaps a few hundred meters from the camp. I lay as still as I could, wondering what would happen next—unready, as yet, to take pusillanimous refuge in the car parked close by. Growing up in Zimbabwe, I'd heard stories of lions and even seen lions fleetingly on visits to game reserves. Lying in the dark under an insubstantial net, the story that was foremost in my mind was the one my father often used to tell.

In 1972 National Parks warden Len Harvey and his new wife were living at a temporary camp at Shapi Pan,[5] close to the center of the 14,500-square-kilometer Hwange National Park. Their house was a rustic, thatched rondavel, with large windows covered in chicken mesh. Shapi Camp had been used as an operational base for the National Parks Management Unit, which at the time was responsible for culling elephants as a population-control measure. The meat from the elephants was dried on wire racks in the camp to make biltong[6] for commercial sale. The abundance of meat inevitably attracted scavenging predators. Lions often ventured into the camp. One of these, a lone lioness, became so bold she even entered unoccupied buildings. The national park staff was used to wild animals in the unfenced camp and paid little heed to the threat of wild animals that lose their innate fear of humans. Late one night, park ranger Willie de Beer was woken by screams. Len's wife, naked and covered in blood, appeared at the door of his house. Despite her hysterical state, Willie gleaned that the lioness had climbed through the window of the couple's hut and was now mauling Len. Willie and his stepson, Colin, armed themselves with hunting rifles and raced to the Harveys' hut. As he peered through the window into the hut, Willie could see signs of struggle. He caught the distinctive smell of the predator. But he could see neither Len nor the lion. He wondered if the lion had already killed Len and dragged his body into the night. In fact, the lioness was crouching right below the windowsill, feeding on

IN MEMORY OF
CYRIL MATHIESON (LEN) HARVEY
KILLED BY LION AT SHAPI
4 TH APRIL 1972
REMEMBERED BY HIS FAMILY & COLLEAGUES

Len's body. As Willie leant forward through the window to get a better view, she struck, grabbing him around the head with her deadly front paws. Her razor-sharp claws sliced his skin to the bone, in the process detaching his scalp so that it covered his face. Willie managed to fire a shot, causing the lioness to release him. Blinded by the blood pouring from his head wound, he stumbled away from the window, followed by the lioness. The enraged animal knocked him over and continued to maul him. When he lay still, she started to lick the blood from his wounded face, a prelude to starting to feed. Colin now confronted the lioness with his rifle. But before he could take a shot, she turned and pulled him down. Unable to see, Willie could hear his stepson's screams and the snarls of the cat. His groping hand found one of the rifles. He used the lioness's growls to orientate himself and pulled the trigger. His bullets silenced the lioness. Though frightfully injured, both men survived. Willie bore the scars of the terrible head wound for the rest of his life. He was awarded the Meritorious Conduct Medal for his bravery

With my pet leopard tortoise, 1973.

that night. The lioness was neither old nor injured, as man-eaters are often supposed to be. A postmortem showed that she had been starving. The desperation of hunger and perhaps an overfamiliarity with humans in the camp she had come to associate with food had driven her to hunt a species she would normally have avoided.[7]

Growing up in Africa, in Zimbabwe, was both unusual and privileged, something I took for granted at the time. My parents' large garden and the surrounding open spaces were paradise for a young naturalist. There was a myriad of small creatures that could be observed or captured— chameleons, lizards, snakes, cane rats, and mongooses. My early interest in zoology manifested itself in the herpetological field. After a friend and I had returned from a frog-collecting expedition in the seasonal wetland near our house, my mother found several frogs disporting themselves in the washing machine—the unfortunate amphibians having been stowed in the pockets of my shorts and forgotten. My pockets were

Zindoga, in my father's research lab.

thereafter thoroughly searched before my clothes were sent to the wash. The highlight of my week was to visit the crocodile pens at the university where my father taught—particularly at feeding time. My friend Luke Madziwa, the technician who looked after the animal house, patiently answered my torrent of questions as he threw chunks of meat to the crocs.

Studying large, dangerous predators is, you might say, "in the family." My dad's research interest was crocodiles, particularly their respiratory physiology but also their conservation in the wild. One of my earliest childhood memories is of visiting the laboratory where he kept some crocs. At the time, Dad and his colleagues were doing some pioneering research to understand how these fascinating but misunderstood creatures function. Wildlife managers still use the drug combinations developed by my dad's team to sedate and handle large crocodilians.[8] I remember visiting the lab when they were calculating the oxygen consumption of a gigantic male croc. The fifteen-foot, half-ton monster

named Zindoga had been recently caught by National Parks rangers in a river pool near a village in the Zambezi Escarpment, where he had been terrorizing the local people. The huge reptile had accounted for the demise of many goats, dogs and, it was said, people. In the lab, Zindoga was quiescent and connected to a respirometer. A giant restrained with ropes and nets, he had to be taken to the Ministry of Transport truck weigh bay in order to measure his poundage. While he slept off the tranquilizing drugs, I got to sit on his back. The knobby, scaly torso of the great dinosaurian beast was large enough for me to sit astride him without my five-year-old feet touching the ground. A few days later, Zindoga escaped his restraints in the middle of an experiment and, with one blow from his massive tail, broke the leg of one of my father's students. The great croc was later taken to the crocodile farm at Victoria Falls, where he was used as a breeding male and became a tourist attraction.

Zindoga.

I saw my first wild lion on a field trip with my father. Field trips had been limited during Rhodesia's civil war in the 1970s. Rural roads were frequently land-mined, and because of the risk of ambush, civilian traffic traveled in convoy with a military escort. After the war, Dad spent a lot of time on Lake Kariba surveying croc populations and searching for crocodile nests on the shoreline and rivers that flowed into the lake. One of my favorite places to go with him was the Sinamwenda Research Station halfway up the lake, which could only be accessed by boat. Getting there took two days in the forty-five-foot research launch, the *Erica*, captained by Ray Palfrey, a former professional hunter from Kenya. The *Erica* had been bought in 1959 for Operation Noah, which saw thousands of wild animals rescued as the newly built Kariba Dam drowned the middle Zambezi Valley.[9] We saw the lion while hiking along the dry, sandy bed of the Mwenda River searching for croc nests

17

Collecting eggs from a wild croc nest at Lake Kariba, October 1972.

around the river pools. We'd been startled by the warning growl of the big cat and glanced up in time to see a flash of tawny fur vanishing into the bush. The sighting was brief but, for a twelve-year-old, exhilarating.

The eggs collected from wild croc nests along the Zambezi and on the shores of Lake Kariba were all placed carefully in a polystyrene cooler box, in a nest of vermiculite, and carefully incubated at the university. Inconveniently, baby crocs hatch from their eggs in late December—during the Christmas holidays when the lab technicians were on vacation. On those holidays, Dad and I used to go to the lab and check the boxes for babies that had hatched overnight. We brought the babies in a bucket to our home, where they swam in the family bathtub. We often had thirty or forty miniature crocs in residence. Care needed to be taken when using the facilities. Hatchling crocs are as ferocious as adults; within minutes of emerging from their egg, they can inflict painful bites with their needle-sharp teeth.

Hatching baby crocodiles was not merely an eccentricity. The idea was that if croc eggs could be collected in the wild and hatched under

Dad with a croc caught during his study at Lake Kariba, 1972.

artificial conditions, there was significant scope for crocodiles to be farmed, rather than hunted for their skins in the wild. Also, a small proportion of the surviving hatchlings could be released back into the wild and, because the natural mortality of croc eggs is extraordinarily high, the released crocs more than compensated the population for the collection of eggs.

In the 1960s crocodile populations had been in drastic decline over much of Africa; many populations were on the brink of extinction. This was largely because of commercial hunting of the animals for their skins, which were valuable for making fashionable leather goods.[10] The sustainable collection of eggs and farming of crocodiles that my father and his colleagues helped put in place turned this around. Leather from farmed animals was of better quality, and along with legislation to protect wild crocodiles, this put the commercial croc hunters out of business. As a result, crocodile populations recovered markedly.

These were the early days of wildlife management. People were becoming aware that wild animals could be sustainably managed rather than eliminated as pests. By establishing markets for their commercial use, incentives were created to conserve wildlife populations. This was a novel departure from the view that wild animals were an impediment to agriculture and must be exterminated to make way for crops or European breeds of domestic livestock. With the aim of developing the nascent wildlife economy, the government established a network of national parks and safari hunting areas across the country. This included the compulsory purchase of former ranch land north of Hwange, which was turned over to wildlife as Matetsi Safari Area. Sustainable use of wildlife, through tourism, game ranching, and sustainable hunting, became the ethos of conservation management.

Lying in the dark with a lion roaring close by, it is easy to imagine the worst. But no lion tore through my thin net that night in the campsite under the teak trees. Eventually, I drifted off to sleep. While there were indeed wild lions nearby, when I eventually returned to the area to start a doctoral study, I discovered the lion I had heard was in all likelihood one kept in an enclosure by the rancher who owned the campsite.

Somewhat unoriginally, the rancher, Skwatula, had named the lion Leo. I was to come to know the pair well over the next few years. Leo was one of many lions Skwatula and his family had raised from cubs, keeping them as enormous and fearsome house pets. Skwatula's house was curiously fortified. During the day, Leo slept in a shady pen, but at night he was released into a fenced enclosure that entirely surrounded the house, to roam at will until morning. Needless to say, the ranch house was never burgled. All the local thieves knew about Leo the lion and the potential consequences of trespassing on Leo's nocturnal watch. Skwatula had great affection for Leo, though the utilitarian nature of their relationship was evidenced by the taxidermied mount of one of Leo's predecessors in Skwatula's living room. Aside from his role as an unconventional security guard, Leo served another purpose. The nocturnal, territorial roars of the captive male, which had alarmed me that night in the campsite, also attracted wild male lions that sought to challenge him for territory. As Skwatula and his family were big-game hunters, the inexhaustible supply of lions attracted to the area was commercially valuable.

Skwatula was a tough, leathery Afrikaner who owned several large ranches, run by his equally rugged sons, in the Gwaai Conservation Area. "The Gwaai," as it is known, consists of several dozen unfenced game ranches and farms whose owners make their livings from hunting and tourism. The Dete Valley (or Dete Vlei, as it is called locally) snakes away to the west, a ribbon of acacia-lined grassland cutting through sandy ridges crowned with teak forest. It leads toward Hwange National Park, the largest protected area in Zimbabwe.[11] Pronounced with a silent H, Hwange, formerly known as Wankie, is in the far west of the country, nestled on the border with Botswana. The park is named after a chief of the Nambya people, a tribe that lived in the northern part of this area in the mid nineteenth century, before they were devastated by the invading Ndebele. Marauding warriors reputedly flayed Chief Hwange alive. Hwange National Park is the easternmost part of the Kalahari Desert; its sandy soils are derived from ancient windblown sand dunes that still form parallel ridges in the south of the park. Wetter conditions have allowed the establishment of forests of teak, *umtchibi*, and silver terminalia that have stabilized the sands. Ancient, dried-up watercourses cut through the fossilized dunes, creating rivers of grass through the dry Kalahari sand. Hwange is part of one of the last wildernesses on earth. One can walk in a straight line westward for 450 kilometers from that small campsite on the Victoria Falls road and not come across a village or farm until one reaches the Okavango Delta in Botswana. The few roads are dirt tracks winding through the bush, following age-old elephant paths. I've had the privilege of working in this wilderness as a wildlife biologist for twenty years.

In the local African language, "Skwatula" translates roughly to "the one who speaks with his fists"—an apt description of his proclivity to punctuate any disagreement with hands the size of hams. He was, in his own way, a gentleman—generous, hospitable, and friendly. I spent many enjoyable afternoons drinking coffee on his front *stoep*,[12] listening to his stories of the Gwaai when it was truly a frontier. I learned a great deal about the area and its history from him. He told me how he had moved there in 1961, after a career as an engineer in the South African Air Force. His family arrived with nothing but a tent, a truck, and a small herd of cattle. The lions soon ate most of the cattle. Cattle ranching was hard on the arid sandveld of western Zimbabwe. It was hot and dry with sparse, poor grazing. Skwatula related how he'd had to send his young sons out in the Land Rover to sleep out in the bush with the cattle herd to protect them. Lions and other predators were not tolerated and shot on sight, as was once common practice on farms and ranches across Africa.

Skwatula was a smart businessman and had pretty soon realized that ranching cattle was not commercially viable, especially if he had to share them with the lions. He turned to the more lucrative livelihoods of game ranching and safari hunting. The area had the potential to support rich populations of wild animals, including elephants, buffalo, eland, impalas, and their predators—hyenas, lions, and leopards. Skwatula and the other ranchers no longer tried to eradicate predators. Instead, they nurtured wildlife for rich foreigners. Trophy hunters would pay handsomely to recreate the bygone safari-hunting adventures of Hemingway and Roosevelt. There were arguably more lions and other wild animals than there had been when cattle ranching was the only form of land use. Some old timers still stubbornly tried to eke out a living by raising livestock, shooting all the predators that came onto their land. But in general, wildlife was seen as a valuable resource, much more so than livestock.

Tourists, who did their shooting with cameras, also came. Hunting was prohibited in the Dete Valley, as it was used exclusively for photographic tourism and run as a concession by Touch the Wild Safaris, owned by charismatic businessman, safari guide, and member of Parliament Alan Elliott. Touch the Wild was jokingly known by locals as "Touch Your Wallet Safaris," as the lodges' rates were mostly too expensive for Zimbabweans. The valley was particularly famous as a wildlife-viewing destination, and one could photograph the "Big Five" (rhinoceros, elephant, lion, leopard, and buffalo) on a single afternoon game drive. It was also a natural funnel for movements of wildlife to and from the national park. Great herds of buffalo and eland and smaller groups of zebras and impalas used the rich grazing and abundant water holes. With herbivores came the predators. Lions, hyenas, leopards, and African wild dogs slipped into the crosshairs of foreigners' Nikons and Hasselblads. This is where I cut my teeth as a wildlife biologist in the 1990s. Inspired by my night at the campsite at the head of Dete Valley I wrote to Alan to ask if I could undertake the field research for my doctorate on the Touch the Wild concession area, to which he generously agreed. I stayed in the guides' quarters of Sable Valley Lodge and Sikumi Tree Lodge. I became good friends with the tourist guides

White rhinoceros cow and calf in the Dete Valley.

Lionel Reynolds at
Makalolo Camp, 1995.

at the lodges and through them learned about the tourism industry and
conservation in general. In particular, the head guide, Lionel Reynolds,
took me under his wing. Lionel was an old-school guide and hunter. To
go on a bush walk with him was to have one's senses completely recali-
brated. He missed nothing. Every animal call and track in the sand had
a meaning. Every creature, from the scurrying dung beetle to the bull
elephant, held his interest. He particularly loved to watch the big cats,
and he passed this fascination on to me.

My first zoological research was not on lions but on jackals, the alert
fox-like canids of the African savannah. I had won a scholarship to
Oxford University in England. David Macdonald, a preeminent carnivore specialist and conservationist, had taken me on as a doctoral
student in the Oxford zoology department's Wildlife Conservation Research Unit, WildCRU for short. David has been an advisor, friend,
and research collaborator ever since. My study of the ecology of African
jackals allowed me to explore the Dete Valley, both on foot and on my

THE VALLEY OF THE LIONS

own in a four-wheel-drive vehicle. I learned how to find and observe the elusive little canids, and how to trap and radio-collar them so I could follow them more easily. Jackals are entirely nocturnal, so my research involved long hours spent outdoors at night. This is a wonderful way to experience the African bush. Many species, especially the predators, are most active during the hours of darkness, when the tourists are gone. If one stays quiet and still, one becomes absorbed into the secret life of the wilderness. There were wild dogs coursing through the long grass, vigilant zebras, and nervy impalas. On bright, moonlit nights, herds of massive elephants drifted past like noiseless shadows. There were lions too. A pride used the valley as their hunting ground. There were several large lionesses with their fluffy, spotted cubs. They would appear for a few days and then vanish for a week or two before reappearing again. Their hunting ranges clearly extended beyond the confines of the valley, so my focus on the lives of the jackals precluded my finding out more for the moment.

Once or twice, big male lions wandered through from the direction of the national park. On one occasion, a coalition of three males spent several weeks in the valley. As part of the jackal study, I was recording jackal territorial calls and playing them back to elicit a response from my study jackals. I'd borrowed a small, powerful speaker from the Oxford zoology department to do this. One slow afternoon, as I had some recordings of male lion roars on an audiotape, I decided to see how the three males would react if I played the recorded roars to them. I set up the speaker in an open area, hidden in a clump of grass a few hundred meters away from the lions. From the speaker I ran thirty meters of cabling to an audiotape player in my research vehicle. I hit the "play" button, releasing a short burst of metallic roars from the hidden speaker. The three males appeared in no time, approaching determinedly, shoulder to shoulder. They were staring intently at the spot where the speaker was hidden, straining to locate the brazen intruder. They had almost marched past the speaker when one of the cats spotted it. He

immediately pounced and seized it in his jaws, punching his huge canines through the thin aluminum. With the others crowding in, he ran
off, head held high, defending his prize from his curious companions.
The cable snaked after him through the grass. The lions were transformed from belligerent territorial defenders to oversized house cats as
they chased the cable. On capturing it, they played tug-of-war until it
was torn into small pieces. When the cats tired of their game and had
moved on, I retrieved the battered speaker, which, though full of large
holes, was astonishingly still functional.

Big male lions seldom stayed long with the Dete Valley pride. Their
wanderings almost always took them beyond the valley. The rumor
amongst the safari guides was that they left the sanctuary of the valley
and were shot by Skwatula's hunting clients—perhaps lured there by
Leo's calls and strategically placed baits consisting of zebra or buffalo
carcasses. There was almost no real information about what really happened to these male lions apart from that they disappeared. If they were
in fact hunted, almost nothing was known about the impact the killing
of all these males had on the lion population. I knew from my under-
graduate zoology studies that male lions are infanticidal; that is, they
kill the cubs of rival males. If the hunted male lions were the fathers
of the cubs in the Dete Valley pride, were the cubs in peril when new
males wandered into the valley? Nobody I asked really seemed to know.
While nobody had an answer, this did not mean people were not concerned; in fact, two men were very concerned.

Lionel, though he was a committed hunter, often spoke about the
constant turnover of male lions and its effect on the lion population.
The other person who was worried was Andy Searle, the tall, lean warden from Hwange National Park. Andy felt that lions were being over-
hunted in the areas around Hwange. As a park ranger in the 1980s, he
had, as part of his job, taken part in elephant-culling operations, so he
was not at all squeamish about hunting. He was, however, a strong advocate for basing wildlife management on sound scientific knowledge.

He felt that too little was known about the impact of hunting lions to justify the current level of hunting. He was pushing for a research study of the lion population and had ensured that the requirement for this research was included in the park's ten-year planning strategy.

David, by then my Oxford doctoral supervisor, was visiting me in the field to check on the progress of my study on jackals and to help out with the research. One evening, Lionel suggested that, because it was a full moon, we all take a night drive to one of the water holes. Armed with coffee, sandwiches, and a hip flask of whiskey, we sat on the ground near Lionel's Land Cruiser listening to the nighttime noises and watching a herd of elephants splashing through the water hole in the silvery light. Into the warm darkness of the African night Lionel put forward an intriguing proposition. Why didn't WildCRU start a scientific research project to understand the effects of hunting male lions? The research would be extremely valuable in making recommendations about how to manage lion trophy hunting and the threats it might pose. Lionel, as a highly respected professional guide and hunter, would support the project. He'd discussed it with Warden Searle, who was enthusiastic and would officially invite WildCRU to undertake the study. So how about it?

As Lionel and Andy had hoped, David and I both saw the opportunity to do a significant and exciting piece of research that could have a great deal of impact on conservation of lions. After I completed my doctoral studies at Oxford, October 1999 found me driving a newly purchased secondhand Land Rover back into the Dete Valley after an absence of two years. David had managed to raise a small funding grant to get the project started, buy a few essential pieces of equipment, and provide me with a small stipend, at least until more funds could be raised to initiate a larger study. I arrived full of optimism, with a pair of binoculars, a camera, and very little in the way of experience studying lions.

# TWO

# STUMPY TAIL

The main lion pride inhabiting the Dete Valley was called the Safari Lodge pride. One of the lionesses in the pride was distinctive. She was large, with the dignity of an African queen. At odds with her proud demeanor, she had a rather unfortunate name. The local safari guides called her Stumpy Tail on account of the fact that she was missing the distinctive black hair tuft that almost all lions have at the end of their tails. I have no idea how she lost it, perhaps in a fight with other lions or a battle with a pack of hyenas. She was also by far the largest female in the pride and, perhaps to make up for her foreshortened tail, she was a lioness with attitude. She ruled the pride and the valley with paws of steel.

Stumpy Tail was decidedly more aggressive than the other lionesses and had something of a reputation with the safari guides for charging furiously at tourist vehicles that were rude enough to get too close to her family. I got a taste of her mercurial temper during the first few weeks of the new lion study. I was watching the pride early one evening, enjoying the spectacle of lion cubs romping in the late afternoon sunshine, lionesses lying indolent as the heat drained out of the day, and the evening light turning to gold. Lions will often move into the open as the sun

sets. They socialize and stretch after a day of inactivity in the shade of a secluded tree or thicket. It is almost as if, as the sun slides out of the sky, they take a moment to contemplate the night ahead, metamorphosing from sluggish butter-colored cats into deadly predators. The lions were relaxed and unperturbed by my Land Rover. Living in a prime tourist area, they were used to vehicles full of noisy tourists, and the sight and smell of people and machines were part of their environment. The curious cubs even approached the vehicle, using its bulk to ambush their siblings in boisterous practice fights. The Land Rover had a roof hatch that allowed access onto the roof rack. It was designed to allow me to climb out of the hatch to take photos or make observations without the disturbance of alighting from the vehicle. The gamboling cubs and evening light made for the perfect photo opportunity, and I duly clambered onto the roof with my camera and sat with legs dangling into the cab, taking in the scene. I was about to learn an important lesson. Lions in photographic tourist areas usually tolerate vehicles and ignore the people in them. This is probably because they only see the outline of

The Safari Lodge
pride, 2000.

the vehicle and treat the whole object and its contents as one inedible and unthreatening entity. They react very differently to the silhouette of a human outside a vehicle. Bipedal creatures have persecuted lions for millennia. Humans, especially those on foot, are seen as a threat by lions and they react accordingly, most often by running away but occasionally by responding aggressively. For Stumpy Tail, the appearance of a human silhouetted against the sky on top of a vehicle in the middle of her relaxing family was too much. Her lazy demeanor changed instantly; her lolling pose became a crouch and her tail lashed backward and forward. Milliseconds later she was on the move, charging toward the vehicle, uttering low, grunting growls. The speed of her movement was utterly breathtaking, and she covered the distance to the vehicle in the blink of an eye. Had she intended to carry through her attack, she could have jumped onto the vehicle with ease. Her purpose, though, was to frighten, not to kill, and she pulled up in a cloud of dust next the vehicle door, growling ferociously. It had all happened so fast that I had barely moved. My camera was still in my hands, unused. To be charged by a lion is an awe-inspiring, adrenaline-boosting experience. The noise and dust, bared teeth, and tensed, muscled body are meant to warn: "I will fight you. I will kill you if I have to. Leave now and leave fast." Almost as quickly as she had charged, she turned her back on me and stalked back to the now-watchful pride. She paused to rub faces with another lioness. The affront to her family had been redressed, the human creature put back in its place. But the mood had changed. Stumpy Tail, without looking back, walked haughtily into the nearby scrub, whether to start the night's hunt or to avoid the annoying human I could not tell. Her low calls from the scrub summoned the rest of the lions to follow, and soon they had all vanished without a trace into the bush, cubs bobbing through the grass in the wake of the adults. I was left to contemplate the empty clearing and the fact that I had a lot to learn about studying lions.

The Safari Lodge pride was spending almost all its time in the Dete Valley then. It was dry and hot, and buffalo and wildebeests were plentiful as there was little grazing away from the valley. At that time of year the water holes along its length were the only water for miles. I spent almost every day observing Stumpy Tail and her pride. Andy Searle had radio-collared two of the lionesses so they could be tracked even when they were in the forests. The pride was sometimes to be found with two enormous black-maned lions. Massive and proud, they were the fathers of the pride's cubs and the lords of the valley. Unimaginatively, they were known as the Black Mane coalition. I got to know the pride well and was able to identify all the lionesses and most of their cubs individually.

To the casual observer, lions often seem indistinguishable from one another. Their plain yellowish coats carry no obvious markings to allow one to tell them apart. Some have unique scars, notches in their ears, or other identifying features, such as Stumpy Tail's missing tail tuft. There is, however, a failsafe way to tell lions apart, and that is by the unique pattern of small, dark spots that surround each whisker. In the

1970s a biologist, Judith Rudnai, studying lions in Nairobi National Park, noticed that the pattern of each lion's vibrissae spots—the small pigmented patches at the base of its whiskers—was different and did not change as the animals matured.[13] They can be used much like fingerprints to distinguish between individual lions. In the Hwange research project, research staff keep a database of ID cards for each animal. We now have a database of more than 700 lions that can be individually identified. This allows each animal in the study population to be monitored so that we can develop an understanding of the demographic and social factors that shape the population.

Including the males, there were nineteen lions in the Safari Lodge pride. Six were adult lionesses of varying ages and eleven were cubs of between six and eight months old. Stumpy Tail was clearly the matriarch and Sinege, whose name translates as "the gentle one," was probably her sister, as they were similar in size and age. The other females were younger and very possibly the adult daughters of Stumpy Tail and Sinege. Amongst the cat family, lions are uniquely social and live in groups, or prides, of related individuals. A lion pride is a sisterhood of mothers, daughters, sisters, and aunts. These familial bonds are the pride's strength. They collaborate in all they do, whether it is pulling down a 1,000-kilogram buffalo, defending each other and the pride's territory against intruders, or caring for the pride's cubs in a communal crèche. Lionesses will seldom tolerate unrelated females in their territory. Lone females caught by an unrelated pride of lionesses are frequently attacked and can be killed or injured. Females alone or in small groups have a hard time of it and are unable to compete with large prides for prime territory. Without the support of the sisterhood, they often struggle to raise their cubs to maturity. As a consequence, reproductive success in small prides is significantly lower than in large prides.[14]

Male lions are no less social than females. They form social groups known as coalitions. Very often the members of coalitions are brothers. The drive to form a coalition is strong. A male lion alone is highly vulnerable to attack by other males, and his chances of claiming a pride and raising cubs are much lower than for a male safely supported by co-alition partners. The urge to form an alliance is so strong that singleton males will form bonds with unrelated males rather than go it alone. This is because it is more advantageous for a male to have the support of a stranger, and share the females in any pride they hold, than to face the hazards of the world by himself. Larger coalitions tend to be composed of closely related individuals.[15] For males already in a coalition with related males, usually brothers or half-brothers, there is no benefit to accepting the presence of unrelated males. Helping raise one's brother's cubs is evolutionarily tolerable, but helping a stranger and protecting cubs that are not one's own makes no sense if the advantage of alliance is unnecessary.

A lion cub's father is critical for its survival, particularly when the cub is still dependent on the pride for food and protection. The gilt-embossed pages of early Christian bestiaries portray lions as fantastical beasts. The legend goes that lion cubs were born inanimate and could only quicken into life when their father breathed it into them. The monks and medie-val scholars who illustrated the manuscripts would, themselves, certainly never have seen a lion and would have relied on second- or third-hand accounts, perhaps garnered from travelers and soldiers returning from the crusades in the Middle East. This implausible myth may have originally derived from the observation that lion cubs are born with their eyes closed and are helpless during their first few days. Without the protection of a territorial male, cubs, even up to the age of two and a half, are vulnerable to being killed by strange males taking over their mother's pride. This callous infanticide seems horrific, but it makes perfect biological sense. In nature, reproductive success for a male lion is to pass on his genes to his offspring and they to theirs, the evolutionary perpetuation of their collec-tive organic image for eternity.

Lionesses will rarely come into estrus and be ready to mate and conceive new cubs until their current litter has either died or survived to around thirty months, the age at which they can survive on their own. However, having won access to a pride, a conquering male does not have time on his side. He is at his physical peak for around four years—starting roughly at the age of five. In that time, he must mate, conceive cubs, and then protect them and the rest of the pride from marauders. He must provide this protection for at least three years to ensure the cubs have a chance of survival. Much can go wrong. Injury, starvation, and death at the hands of other lions, people, or dangerous prey animals are everyday realities for all big cats. This is why male lions are impatient and do not wait until the previous male's cubs reach maturity. Waiting would mean a delay in conceiving his own cubs for several years, perhaps missing the opportunity to raise cubs at all.

Having wrested control of a pride from a rival, a male lion promptly dispatches his vanquished rival's cubs. A quick bite through the spine and a shake is all it takes. The fluffy bundles, once joyously energetic, lie scattered in the bush—punctured by massive canines. Food for hungry scavengers. The heartbreaking aftermath of a pride takeover is not an image that is easy to forget. Yet this is how nature works. The teeth and claws of a newly dominant male are invariably red with the blood of his rival's children.

Yet as most field biologists quickly discover, there are no hard-and-fast rules in nature. Behavior in intelligent, social species is infinitely variable, and the reality does not always conform to what evolutionary theorists would predict. At the demise of their former pride male, lionesses do not lightly accept the death sentence imposed by biology upon their babies. Sometimes they will furiously confront new males with flashing teeth and raking claws, which may be enough to intimidate newcomers. But lionesses are significantly smaller than adult males, and in a one-on-one confrontation, they will certainly lose. The project has recorded several incidences where lionesses have fought to the death in a brave but futile defense of their cubs. Sometimes, rather than fight and risk injury, a male will lay siege to the female and her cubs. The lioness

and her babies have nowhere to go. It becomes a waiting game, and the males are patient and vigilant. One moment of inattention from the mother and it is all over. At other times, lionesses will flee with their cubs to hide from the would-be killer. Sometimes this fugitive strategy works; the new male loses interest and moves on. More often, once on the trail of the female and cubs, he is relentless in hunting them down; the sad outcome is inevitable.

However, lionesses have more subtle strategies in their behavioral armory. They will often use all of their feminine guile to bewitch and confuse the new and foolish male. If they do not fight or flee, lionesses turn the male's evolutionary drive to mate into his biggest weakness.

Feigning submission to the new male, a lioness becomes slinky and
sinuous, oozing charm and sex appeal, tail curling snakelike across the
male's head and shoulders as she entices him to follow her. By mating
with him, she distracts him and allows the other lionesses to escape
with the pride's cubs, including her own. By intoxicating him with feline charm, she wins time for her children.

The two magnificent males in the Black Mane coalition were almost
identical to look at and were most likely brothers. They had luxuriant
black-brown manes that covered their heads and grew down onto their
shoulders like thick, woolly capes. Male lions from the dry Kalahari, of
which Hwange is a part, are famous for their long, dark manes. This may

Radio-tracking lions with a
receiver and antenna.

be because the Kalahari nights are freezing cold in winter and the arid climate is never particularly humid, so having a thick mane is an advantage rather than a burden. The Black Mane brothers were often with the pride, but sometimes they would vanish for several days, the business of territorial defense or extramarital dalliances taking them away from their demanding family life with the Safari Lodge pride. These excursions took the brothers beyond the valley, and I was curious about where they went and what they did in their travels.

As one of the males was fitted with a radio collar I was able to follow them. In 1999 the cutting edge of animal telemetry was a very high frequency (VHF) radio transmitter mounted on a collar. The radios transmitted a signal every few seconds, and with the aid of a receiver and a directional antenna, these "bleeps" could be tracked over a distance of around five kilometers. This technology was hugely useful in following the movements of the lions, but at the same time, it had its limitations. It soon turned out that lions move vast distances— sometimes covering ten to twenty kilometers a night. I was effectively limited to following them in a vehicle from roads and dirt tracks. So when they would sometimes disappear into areas inaccessible by road, I would spend several frustrating days searching for them. Later I bought and learned to fly a microlight aircraft. From the air, I could detect the collar's signals from up to twenty kilometers away, and this greatly increased my ability to keep track of the lions. Simple radio collars almost became obsolete a few years later with the advent of collars fitted with global positioning system (GPS) loggers. Using the same technology as vehicle satellite navigation systems, these collars logged the lions' exact position on a preprogrammed schedule, with later models transmitting this information to researchers via radio or satellite link. When the project started using GPS collars, the huge volume of superbly accurate information on the lions' movements made me realize just how far many of them were ranging. Some males covered ranges of more than 1,000 square kilometers—an area roughly a third larger than New York City or two-thirds the size of Greater London. This was to have significant implications for our understanding of how even seemingly vast protected areas might be inadequate to protect such wide-ranging predators.

One morning, I found the pride and their cubs settled for the day in a grove of acacia trees at the western end of the valley. The males were not with them, and I was unable to locate a signal from the collared male's transmitter. They had clearly departed on another of their forays, and I was determined to discover where they had gone in order to get some insight into their behavior at times when they temporarily left the pride unprotected. They had obviously walked a considerable distance during the night, as I could not get a signal from them, even from various high points in the vicinity. After several hours of my searching, the usual nagging doubt emerged. Had the radio collar malfunctioned? Had something happened to the collared male? Surely they could not have vanished so completely.

It was starting to get hot and I was about to abandon the search when I tried once last time from a small ridge at the side of the Dete Valley. Standing on top of the vehicle with the antenna gave me a little extra range. The faint *pip . . . pip . . . pip . . .* from the radio receiver informed me that the collared lion was far to the west, across the railway line that demarcated the boundary of Hwange National Park, at the far limits of the radio transmitter's range. I had not realized the males moved so far that way, and I excitedly headed in the direction of the signal to find out exactly where the two lions had gone. This distant location would give me some idea of how large a range these lions covered—a crucial piece of information in understanding how they should best be protected. The signal became progressively stronger until it became loud enough for me to be sure the lions were somewhere close by. Eventually I narrowed down the position to a large patch of spiny hook thorn acacia scrub. To drive a vehicle into this kind of country is not generally advisable. Iron-hard thorns from the Chinese lantern bushes wait to rip open tires. Aardvark holes can collapse, miring a vehicle in the sand up to its axle. Both situations would necessitate alighting from the vehicle to change a tire or jack the vehicle out of a hole, something I did not fancy with two large male lions in the close vicinity.

Curiosity, however, got the better of me. After a day of searching, I wanted to see the lions and at least record a behavioral observation to justify the time I had spent in the search. I edged the Land Rover into the scrub, squeezing between thornbushes whose branches clutched at the vehicle's sides. A few hundred meters into the scrub, the terrain opened up into a large, grassy clearing. The signal was now so strong that the sound was distorted by the radio receiver's small speaker. The lions were clearly within twenty or thirty meters. Still, they were not visible in the thick vegetation surrounding the clearing. With binoculars I scanned the surrounding bush. A black-tipped ear flicked in the dappled shade of a hook thorn bush. Between grass stems and fallen brush, large, watchful eyes came into focus. The shape of the big cat slowly appeared. Behind him the second male came into view. The giant cats were curled up together like two puppies. They lay almost touching, taking comfort from one another's presence. I realized then how closely bonded are the partners in lion coalitions, especially those who have been together since birth. Even though they are the biggest and most fearsome predator on the African savannah, their reliance on their coalition companion is total. Without the support of brother and ally, tenure of a pride is short and insecure; defeat by a stronger coalition is inevitable.

The members of the Black Mane coalition were doing what all male lions are hardwired to do: defend the piece of real estate they have commandeered. Male lions regularly patrol the boundaries of their territory and advertise their presence with scent and sound. They see off any potential intruders. A patrolling male does a circuit along roads and paths that define his territory, stopping every so often to spray-mark a bush or tree at the side of the track. He does this by backing up to the bush

Males in a coalition form strong bonds. They can often be found resting in close contact with each other.

and arching his tail across his back, and in seeming defiance of male morphology, he sends a spray of urine backward. This odiferous signpost provides his rivals with the olfactory information that he is active in the territory. To underline his message, a male lion also roars regularly. The low-frequency sound reverberates across the land and can be heard up to ten kilometers away. Lions will often call in the coolest part of the night or early morning. At these times, the cool air is denser, and this is thought to propagate the sound more effectively. Lionesses will also roar, joining together in a resonant bass singsong. As lionesses are also vigilantly territorial and will engage in brutal territorial combat with rival female prides, these roaring sessions also serve to advertise their ownership without the need to fight. Actually, it turns out that lions can, by ear, count the number of rivals calling and assess their chances of winning a territorial contest. Two or more adult females will confront a single lioness from a rival pride but avoid tangling with a pride of the same or larger size.[16] Lions are not stupid. Picking a fight with a gang of neighborhood bullies, especially those armed with pointy claws and teeth, can get you badly hurt or worse.

The Dete Valley was a haven for wildlife and lions, but it was also an island surrounded by trophy-hunting concessions. Each time the pride left the valley, they were at risk, especially the big males, who were prime targets for hunters—though hunters would sometimes shoot females too. It was always a vague worry when the pride left the valley. But after nearly a year, none had come to any harm, so I'd stopped worrying about it.

One day in June 2000, I got a message over the camp two-way radio. As is often the case in rural Africa, the message was passed on verbally and often through several people, resulting in embellishments and omissions.

"Rob has some collared lions on his property. Can you come over to see them?" came the radio call. "Copy. Over."

"Yes, copied that," I replied. "Tell him I'll come over tomorrow."

Rob was the manager of a nearby private hunting concession known as Antoinette, on the border of Hwange National Park. He could only be calling about the Safari Lodge pride, as they were the only lions we had radio-collared at the time. The Antoinette concession was directly adjacent to the park boundary and separated only by the railway line that demarcated the northern border of the protected area. It was a long way beyond what I had thought was the pride's normal range, about twenty kilometers from the Dete Valley as the crow flies. As I had not seen or been able to find the Safari Lodge pride for several weeks, I was pleased to get news of them.

I duly drove over the next day to try and catch a glimpse of them, stopping for more information at the dilapidated farmhouse that served as the safari headquarters and home to Rob's family. I knocked on the screen door. Rob, with excited hunting dogs swirling around him, opened the door. He somewhat sheepishly invited me in. My look of inquiry prompted him to gesture to the kitchen table. My heart sank as I saw the skulls of a lion and lioness displayed side by side, cleaned of flesh, teeth bared in an everlasting snarl. My stunned silence must have jarred Rob.

"Sorry, thought you got the message," he blurted out. "We had a client in for a lion hunt this week. We didn't see the collar with all that mane. Really sorry, mate."

Feeling dazed, I picked up the lioness's skull.

"Hunter was a Spanish chap," Rob mumbled. "They often pay to shoot a male and female to display together in their trophy collection. We found these two under a thorn bush on the concession. Probably a mating pair. Funny, though, when we shot the male, that bitch charged us instead of running away. I had to put a bullet through her chest to stop her. Was pretty close. She could have scratched us up bad."

Off-handedly, he added, "She's not even a very good trophy. She was missing the end of her tail."

This was devastating news. The loss of Stumpy Tail, the pride leader, and one of the Black Mane coalition, was a disaster for the pride. What would happen to them now that two of their protectors were gone? They were nowhere to be found; I hoped they had headed back to the safety of the Dete Valley. Rob was basically a decent guy, and later, over a cup of tea, he explained that he did not like having to hunt lions.

"I hate hunting cats, but it's the only way this hunting concession is economically viable, and the South African owners insist we make money."

"I'd lose my job if I didn't do it," he added, looking embarrassed.

There is no doubt that the financial incentives to hunt lions were and are huge. Lion hunts in prime areas are marketed to wealthy foreign hunters at between $50,000 and $100,000 a hunt.

There was no resident pride of lions on Antoinette,[17] but it was virtually certain, because of the geography of the area and the close proximity of the national park, that lions would pass through sooner or later. The farm is on the northern end of a long valley, the Masumamalisa Vlei, that cuts southward through the teak forest deep into the national park. Like the Dete Valley, it is teaming with wildlife, which in turn attracts predators. This topography makes Antoinette a natural trap, perfect for attracting lions. The cats amble up the valley from the park and are effectively funneled toward Antoinette on the park boundary. What is more, a strategically placed bait, in the form of a zebra or buffalo carcass hung in a tree across the railway line, is guaranteed to lure the hungry cats out of the sanctuary of the park.

Hunting lions on the Hwange National Park boundary, on small hunting concessions like Antoinette, had been going on for decades. What had happened to Stumpy Tail and the Black Mane male was not illegal. The National Parks Authority had approved the quotas and the property owners were perfectly within their rights to sell the hunts to a foreign trophy hunter. Many argued that hunting was providing much-needed revenue to the country and a justification for setting aside hunting areas for wildlife, rather than turning them into cattle ranches. Yet even after just over a year of monitoring the lions, it was clear that trophy hunting of lions had the potential to have a huge impact on the population. Hunting lions on small properties like Antoinette was impacting lion prides that ranged much further afield. This was extremely worrying, but understanding the impact and helping wildlife managers formulate policies to better protect lions was also the reason the Hwange Lion Project had been set up. Incidents like the hunting of Stumpy Tail and her companion all fed into our understanding of how hunting affected Hwange's lions. It was hard, though, not to focus on the individuals and feel a real sadness for the animals whose characters I had come to know so well. It seemed a senseless waste to have killed these beautiful creatures. I'd seen how social these lions were and how dependent they were on each other. The loss of key individuals in the pride must surely have an impact on the lives of those lions remaining. Stumpy Tail was a fearless hunter and mother whose experience helped to keep the pride and its young cubs safe and fed. The pride would be weaker and less able to function without her to guide them.

Killing key individuals from a population of social animals, such as killer whales or elephants, is known to have harmful consequential effects that disadvantage or cause the death of other surviving individuals in the social group. Lions are no less social than whales or elephants. Would the same apply when a key individual was killed? In the early days of the lion study, these questions were starting to arise in my mind. Questions that would challenge the prevailing wisdom that trophy hunting, or "sustainable use," was beneficial to the conservation of populations of lions and other wild animals.

Anthropomorphizing animal behavior is frowned upon in scientific circles. Young zoologists are taught to interpret the behavior of their subjects in the sterile terms of evolutionary theory—acquisition of food and mates, defense of resources. Yet when I heard the remaining Black Mane male calling for his missing brother for days on end, it was hard not to interpret this behavior as a distressing expression of loss.

# THREE

# WELL HELL BILL AND THE FLYING BRICK

The most effective way to study the movements and behavior of lions in the wild is through radiotelemetry. Using this technology, you can locate a specific cat and track its movements, even across a massive range of rough terrain or thick bush. How do you go about doing this? First: catch your lion. This is where theory and practicality part company. Catching the jackals for my doctoral study had been tricky enough. Jackals are clever, suspicious little creatures that don't readily allow humans to approach them. To capture and fit radio collars to them, I'd had to get almost as wily. I'd use small pieces of meat as bait to lure them and, once they were used to the baits, set humane foothold traps to catch them. The little critters were incredibly smart. If they detected a whiff of human scent on the trap or around it, they wouldn't come anywhere near. I'd check the traps every few hours; if I'd caught a jackal, things became interesting. I'd leap out of the car and throw a blanket over the struggling animal. Then I'd grab its head, avoiding the snapping jaws and needle-sharp teeth. Jackals can carry rabies. Part of the reason I was doing the research was to understand potential transmission of the deadly disease between jackals. A bite from a trapped jackal would mean a trip to the hospital and a course of prophylactic rabies injections. Luckily, I was never bitten. Once overpowered and bundled into a blanket, the little canid would stop struggling. I'd fit the radio collar, put the jackal in a sack, and weigh him with a fisherman's spring balance. Then I'd release him. All within a few minutes.

Occasionally a lion or leopard would investigate the traps and sometimes set them off. But the traps were too small and light to catch their feet, which was lucky, because I don't know how I'd have released an angry lion from a trap. The blanket and rugby tackle method would certainly have ended in disaster. I did, however, get a taste for what it might have been like had I caught a lion. Early one morning I went to check the traps I'd set the night before. I was accompanied by a young British student who had come to work for a few months as a research assistant. This was his first day in the African bush. On approaching the trap, I was immediately alarmed by the ferocious growling sounds I could hear and, soon after, by the sight of an angry spotty creature. At first I thought I'd caught a leopard, the fiercest and, for its size, the most dangerous of all the predators. How was I going to solve this problem? My initial panic eased somewhat when I realized it was not a leopard but a three-quarter-grown spotted hyena. This was only slightly less alarming. The hyena's powerful, vice-like jaws are adapted for cracking open and crushing the large herbivore bones they eat. A determined bite from a hyena would certainly break an arm or leg. What to do? I had to release the animal, but I had no way to tranquilize it. After a few moments' thought, I decided to try the jackal-catching technique. I tossed the blanket over its head and leapt on. It was like riding a bucking bronco. I knew hyenas were superbly adapted to lifting and dragging large carcasses and as a result have incredibly strong necks and forequarters. I just had no idea they were this strong. I got a good grip of its jaws, but it took both hands and all my strength to hold them closed. My full weight was on its neck and shoulders, but it was still lifting me off the ground. While I had at least temporary control of the business end of the beast, there was no way I could let go to release the trap mechanism.

Catching a jackal to radio-collar for my doctoral study.

I needed help. My gaze fell on the student. His eyes were as wide as saucers as he watched a lunatic wrestling with a hyena. He obviously wondered what the hell he'd signed up for. I shouted to him to get out of the vehicle and release the hyena's foot. He was understandably hesitant, given that the hyena was growling savagely and I obviously barely had it under control. Eventually he plucked up his courage, scuttled out of the vehicle, and released the spring that held the raging animal's foot. Now for the dismount. I let go of the hyena's chomping jaws, simultaneously leaping back as far as I could. In the process, I tripped and fell over backward. Fortunately the hyena, assessing its immediate options, chose freedom over revenge and galloped off into the bush as fast as it could. Whew!

Catching lions could not be more different. They most definitely need to be tranquilized before they are handled. This involves using a tranquilizer dart gun to fire a light plastic or aluminum syringe dart at the cat. The dart is filled with a cocktail of anesthetics and tranquilizers designed to knock the lion out for long enough to handle the animal, take measurements and samples and, if necessary, fit a radio collar. The drugs take about a quarter of an hour to take effect; thereafter it is prudent to check whether the cat is fully immobilized. Sometimes it has not received a large enough dose of tranquilizer or has not yet succumbed to its effects. One should cautiously approach the recumbent cat and give it a gentle prod. If it does not respond, a slight tug of its tail is a useful confirmation that it is safe to handle. Because lions almost always have curious friends, it is wise to post a lookout to make sure others in the pride do not approach too closely.

My first lion capture was with Andy Searle. At that stage I had not had the necessary training to handle the drugs and the dart rifle myself. The lions we were after were at a place called Kapula in the north of the park. The pride there had been seen feeding on a kill—the perfect opportunity to capture and collar a couple of the cats. I watched as, with needle and syringe, Andy punched through the rubber caps of the drug bottles and deftly drew up and carefully measured out dosages. As he filled the darts, he explained what each drug was for.

"This drug is called Zoletil," he said. "It is a dissociative anesthetic; it causes a temporary paralysis. If you inject yourself by accident, you'll take a trip to the moon and back.

"You have to be careful when using this drug—the lion can still see, hear, and feel pain. Despite the immobilizing effect of the drug, they'll wake and get up if you stimulate them. So we add a sedative to the mixture too. We'll hand-inject the antidote once we have finished with the procedure."

With the syringes filled, he loaded the first into his dart rifle. "OK, let's go," he said. We jumped into the Land Rover and drove toward the lions.

"We'll dart the big female and the young male," he announced.

Moments later, with a dull *phup* from the dart rifle, he fired off the first dart. It thwacked into the female's rump. She leapt up, snarling in surprise, indignantly swatted the lion feeding next to her, and went back to eating. Soon the male was darted too.

Handling lions for the first time is an awesome experience. Being right next to them, you get a sense of just how big they really are. Their muscular bodies are covered in short, coarse fur the color of buttermilk—smooth and sleek, almost silky when you stroke it. The oils on their skin, which quickly transfer to your fingers as you handle the animals, have a distinctive smell that is not unpleasant, almost sweet. Working with lions, even those that are safely tranquillized, is not without danger.

"Need to wash our hands carefully after this." Andy said. "Lions carry nasty tapeworms, and if you ingest the eggs, they can form unpleasant cysts the size of a grapefruit on your internal organs."

Andy had been in National Parks for two decades and was a hugely experienced wildlife and conservation professional. I learned an enormous amount from working with him. A fervent, hands-on conservationist, he was one of a breed of men and women who dedicate their lives to a cause, benefitting little in monetary terms. In those days, a National Parks salary, even for a senior officer, was a pittance and barely put food on his family's table. He had almost no budget to run the National Parks Management Unit. Everything was done on a shoestring, largely from donations from overseas conservation charities. The plan was that I would work closely with him on the lion research project. I would also move to Umtshibi, the Hwange headquarters of the management unit, where a small cottage was being converted for my use. Andy's hope was to establish a nucleus of research expertise that would address pressing wildlife-management problems. There were already a few other researchers based there. Greg studied African wild dogs. Viv Wilson and his field assistants, Julia and Marion, were undertaking research on spotted hyenas.

As the head of the park's management unit, Andy was also responsible for managing and monitoring the rhinoceros population in the park. At the time, there were between ten and twenty white rhinos and about a hundred black rhinos in the hilly areas to the north of the park. Rhinos are heavily threatened by poachers. Their horns, used in traditional Chinese medicines, are, weight for weight, worth more than gold. For this reason, they have been hunted to extinction over most of Africa. Andy had radio-collared many of the white rhinos so they could be monitored and more closely protected. He would regularly track them from a small two-seater Robinson R22 helicopter. Many flying hours in challenging conditions on rhino and wildlife management operations meant he was a hugely experienced pilot. He had once even transported a lost rhino calf, suitably sedated, in the passenger seat of the helicopter to reunite it with its mother.

Andy, the veteran wildlife professional, made catching lions seem easy. The first lion capture I did after qualifying for my capture license went a lot less smoothly. One evening in January 2001, at Mpofu Pan, a water hole just outside the park boundary, I found four burly young

Andy Searle flying a rhino calf to be reunited with its mother.

males feeding on a buffalo bull they had killed less than an hour before. This was the perfect opportunity to dart and collar these lions. I guessed they were unlikely to move far from the buffalo carcass and, unless disturbed, they would feed all night. I planned to return before dawn the next morning and went home to collect the darting equipment and organize some people to help. At first light the next morning, I was back at Mpofu Pan, accompanied by Paul (at whose tourist lodge I rented a small cottage), Malaki (the lodge's safari guide), and Lindy (an American biologist temporarily helping me with fieldwork).

Sure enough, the lions were still there, gorging themselves on the buffalo despite the fact that their stomachs were already distended. They were so focused on feeding that I was able to slip within ten meters of them. Soon I'd darted two of the cats. They barely budged; no doubt they were less worried about the pinprick in their well-muscled behinds than relinquishing a tasty piece of buffalo. Both were soon sound asleep with their noses firmly in their dinner. I drove slowly up to the carcass to persuade the other two to move off, which they did reluctantly, settling down under a small tree to observe the proceedings. I blocked their view of the carcass and their prostrate buddies with the vehicle and we went to work on bolting on the collars. Care needs to be taken when handling immobilized lions. In effect, the anesthetic drugs paralyze the lion temporarily, while a sedative in the tranquilizer cocktail causes relaxation closely akin to sleep. However, the lions can quickly rouse if handled roughly or subjected to any painful procedure. This becomes a lot more likely as the drugs are metabolized and their effect begins to wear off. The previous year Andy had received a painful swat from a too-lightly tranquilized lion as he punched a tag through its ear. While he held the groggy animal down with the one hand, he'd had to ask an assistant to extract the half-awake lion's claws that were embedded in the back of his other hand.

We fitted a collar to the larger male and I fixed a small yellow ear tag through each lion's ear as a semipermanent identifier. The collaring operation had gone smoothly and we congratulated ourselves on a job well done. The sun was now getting higher in the sky and the day was getting hot. It was nearly time to inject each lion with the antidote and leave them to recover in peace. I was worried that the recumbent lions were in the direct sun. The sedative drug can interfere with their ability to regulate their body temperature; they could dangerously overheat before they had revived. We decided, because they both appeared well anesthetized, that we would move them a few meters into the shade of a small bush. It would take all four of us to shift the drugged cats, each of which must have weighed between 180 to 200 kilograms. Rather than try to lift them, our plan was to roll each cat onto a small tarpaulin and, using it as a makeshift sled, drag each lion into the shade. Because of their heavily distended bellies, we had to roll them over onto their stomachs to avoid the risk of twisting their intestines. This strategy worked well for the lion closest to the bush, and he was soon installed in the shade. We returned for the second lion. It was now more than an hour since I had first darted the pair, but they still both seemed well sedated. We set about rolling the second lion onto the tarpaulin, with my helpers wisely assigning me the head end. I supported his huge cranium in the crook of my arm while gripping his new collar and a handful of mane with my other hand. I was so close to him that I could feel his moist, buffalo-scented breath on my face. As we rolled him onto his belly and chest, I felt his body stiffen. I was no longer taking the full weight of his head. Rather, with one arm round his neck, I was looking, from a foot away, into the yellow-golden eyes of a very awake lion. A lion that was, in turn, starting to take an interest in my face. We froze in this pose for a heartbeat or two before I yelped a warning to the others and hurriedly let him go. In the ensuing chaos, humans and lion exploded in all directions. The big cat stumbled unsteadily to his feet, galloped off a few paces only to trip over and collapse onto his sleeping brother, who in turn awoke in fright and clumsily staggered to his feet. Of all of us, Lindy chose the most sensible option and quickly scrambled up the ladder onto the Land Rover roof. Malaki, finding the ladder occupied by the nimble Lindy, slid under the vehicle and lay facedown in the dust, imagining that his end had come. Paul and I both chose the sparse sanctuary of the same small bush—colliding and crumbling into a heap. The diaphanous bush barely shielded us from the two lions. Paul erratically waved an ancient World War One Webley revolver he had somehow produced from inside his overalls. The now-wide-awake lions took full advantage of the confusion, loping off unsteadily to retire into the safety of a nearby thicket. I don't know who had been more terrified—the lions or us. As the adrenaline wore off a few minutes later, the ludicrousness of the situation dawned on us and we could not stop laughing. No doubt the lions were also having a laugh at the expense of the incompetent humans. We named the new lions the Mpofu coalition after the water hole at which I first darted them. They were to become one of the core groups of lions in the study.

. . .

Since that first lion capture, I've lost count of the number of lions I've darted and handled. Every capture is different—some dangerous, some chaotic, none of them boring. However, fitting a collar onto a lion is only the beginning. Thereafter, in order for any information to be gleaned about the lion's movement and behavior, it needs to be tracked. Lions are hardly ever obliging enough to stay on roads; they sometimes disappear for days into the thick bush and scrub. The best way to locate them is from the air, as the added height greatly increases the distance over which the collar's radio signal can be detected. I soon realized that I'd need to get airborne. The most cost-effective solution was to buy and learn to fly a microlight aircraft. These are two-seater aircraft of less than 500 kilo-grams. They are great fun to fly and perfect for tracking animals, because they are slow and can be flown relatively safely at low altitudes. My first microlight was a locally made aircraft called a Fulcrum, with registration number Z-MED. It was basically an aluminum frame with Dacron sail-cloth stretched over it and a 100-horsepower petrol engine driving the propeller up front. As my flying instructor noted comfortingly, "Better hope it never catches fire, because everything on this plane is flammable."

A few years after getting the plane, I brought in a volunteer pilot to help me with the flying. He was a laid-back Alaskan, nicknamed Well Hell Bill, as he prefaced most conversations with a drawn-out "Well, hell . . ." The first time Bill flew Z-MED, he taxied up after his flight, cut the engine, and drawled, "Well, hell, Andy. She handles like a flying brick, but I reckon we'll manage." Thereafter Z-MED was christened the Flying Brick. Bill and the Brick did many hours flying together while we tracked lions.

I used to keep the plane at the main commercial airport, just outside the park, in some disused air force hangars left over from the Rhodesian Bush War. The whole airport was in mothballs, as there were no longer any scheduled flights to Hwange. The tourist boom of the 1990s had collapsed with the uncertain politics in the country. Most of the time

Z-MED, the Flying Brick.

the microlight was the only aircraft using the facility. The airport's runway was nearly five kilometers long—complete overkill for a microlight that, with flaps fully deployed, only needed 100 meters to land. Hamlet, the air traffic controller, and his staff were so bored that every time the little plane came in they'd turn out the fire engine, the ambulance, and airport security, just as if a 747 was landing.

Nothing is more beautiful than floating over the Kalahari bush in the early morning in a tiny airplane. The horizon seems infinite, sky and land merging somewhere in the distant haze. There is something mystical about being alone in the African sky at dawn. There is also nothing more terrifying than fighting to control a bucking microlight when the wind picks up strength and the turbulent thermals buffet the tiny machine. At those times I always thought of the article reporting that, among biologists and conservation professionals who died on the job, 66 percent did so in aircraft accidents.[18]

Hwange is huge and remote—a wildlife refuge with few people and great empty spaces. That's the point, really. It is a piece of virgin Africa

where ecological processes can follow their centuries-old patterns without the destructive intervention of humans. You get to know the animal inhabitants around camps and bush houses in which you stay.

The project base camp was often visited by a pack of banded mongooses, twittering and scuffling through the dust. Anything edible left within reach in the kitchen was always quickly demolished. A hive of bees lived in the wall cavity of one of the bedroom huts. On a hot day, the small building sounded like it was about to take off as the bees fanned their honeycombs to stop them from melting. The scratching sound we kept hearing in the ceiling of the office turned out to be a large cobra that was hunting the bats that roosted there.

Before we were married, Joanne and I lived for a time in a tiny cottage in a grove of *umtshibi* trees at Dynamite Pan, eight kilometers from the main park headquarters. The rustic two-room building was baking hot in summer and freezing in winter. Hot water was obtained by lighting a fire in an ancient wood-fired donkey boiler that belched smoke and steam in equal measure. We shared the cottage with a family of dormice, who ate holes in our clothes. Bush babies leapt with wild abandon through the trees at night. Fat geckos sat under the eaves munching on insects attracted to the electric lights. We loved the secluded spot and being immersed in the life of the park's animals.

Joanne and I had met as students at Oxford, and she left her job in a financial investment firm so we could spend time in Africa together. On her first day in the park, a young elephant bull we had stopped to watch trumpeted and flapped his ears at us in teenage temper. In fright and needing to feel separated from the unfamiliar, Joanne rolled up the Land Cruiser's window. She soon warmed to the adventure of bush life. After a week or two she was as unfazed by Africa's critters—dangerous and benign, crawly, bitey, or poisonous—as those of us who had grown up with them.

A drive through the park will almost always provide sightings of some of the larger inhabitants. Gawky giraffes pruning the gardenia bushes into fantastical topiaries. Herds of buffalo churning up dust clouds that the sunset paints a fiery crimson. Elephants are everywhere. Hwange has one of the largest populations of bush elephants of any reserve in Africa. Most of the time, elephants ignore vehicles; they know cars pose no threat. It is easy to get blasé around the huge beasts; they become part of the daily landscape, a different kind of traffic to be negotiated on the road. But it does not do to take these massive animals for granted. An angry elephant can overturn and smash up a vehicle with ease. I once saw a tour vehicle that had been given the elephant treatment. An elephant cow had materialized out of the bush and, with no provocation, attacked. The cab had been completely flattened; its metal sides, where she had repeatedly driven her tusks, looked like a colander. Luckily the group of gap-year students had managed to escape out the opposite side of the vehicle and had watched, appalled, from behind a tiny bush as the enraged elephant systematically dismantled the truck.

One day, I was bumping along a sandy track in the park. Rounding a corner, a small herd of elephant cows and calves came into view on

the road fifty meters ahead. They seemed agitated. I stopped the vehicle and switched off the engine to give them space and allow them to move off. Give elephants respect and, generally, they will not bother you. Not this time. The matriarch trumpeted, throwing up dust. Mrs. Pachyderm was annoyed. She picked up a large tree trunk and threw it in the air as effortlessly as a child throwing a pencil. The next minute she was charging down the road toward our vehicle, squealing like an enraged freight train. Mrs. P was seriously pissed off. Time to leave.

I turned the ignition key. The starter motor tried to engage. *Click, click, click,* silence. The engine would not start. The battery terminal, which kept bumping loose on the rough roads, had required some attention that morning. But now was not the time to fix it. Mrs. P was

Joanne (left) and Jane (right)
radio-collaring Mpofu.

gaining momentum. Head down, trunk curled up under her chin, she meant business. The front bumper of our vehicle was firmly in her sights. I tried the ignition again. *Click, click, click,* silence. Asa, the Swedish veterinarian who was helping me catch some lions, gripped the dashboard, eyes wide with alarm. I turned the ignition key again. *Click, click, click,* silence. Mrs. P was now just a few meters away. Asa and I held on tight, bracing for the impending impact. Mrs. P pulled up, towering over us in a cloud of dust. She squealed angrily and shook her massive head and ears, spattering the vehicle with dry mud. She stared down her trunk, as if we might owe her money, her tusks pointed at the windscreen like great ivory can openers. We sat frozen, not daring to move. Then, abruptly, she ambled off into the bush and started feeding on a nearby terminalia bush. We waited quietly for a few minutes, hearts pounding. Oddly, she now seemed completely relaxed. I tried the ignition again. *Click, click . . .* the engine kicked into life. I gingerly backed up the road. Mrs. P blithely carried on feeding.

The people who choose to live and work in remote wildlife areas are often colorful and unconventional characters. Some are down-to-earth, practical people. Others are running away from life's realities. Many of these bush people work in conservation, as hunters, or in the tourist industry at local lodges. Others are just plain eccentric, preferring the solitude of the bush to the hustle of city life. One of these was my friend John Foster, who lived as a virtual recluse in a ramshackle farmhouse just outside the park. He was a retired telecommunications engineer with a long white beard, which he kept tidy by tying it up with pieces of multicolored telephone wire. John hated to sleep indoors, so he moved his bed onto the roof of the house and from his perch would wave, stark naked, to the tourists on their early morning game drives. John seldom bothered with a shirt but instead wore a shoulder holster cradling a long-barreled .44 Smith & Wesson. If you asked him, he'd show you the dents left on the revolver where the weapon had stopped the scything horn of the charging buffalo that would otherwise have skewered him through the chest.

At the more conventional end of the bush people scale was Jane Hunt, or Jungle Jane, as she is known to most people in Hwange. Jane used to work as a professional safari guide at high-end safari lodges across the country. She was one of the first women to pass the stringent guide's exams and qualify as a Zimbabwean professional guide. When I met her in 2002, she'd run her own horseback safaris and managed safari camps all over southern Africa. By then she was tired of looking after tourists and was much more interested in looking after lions. She joined the project in 2003 and probably knows more about Hwange and its lions than anyone else alive. Apart from her competence as a field researcher, Jane is an inveterate collector of zoological artifacts she finds in the field. The small cottage she lives in is adorned with collections of antelope horns, small animal skulls, and elephant patella bones, which are artfully arranged amongst the pot plants. Okay, she's not so conventional after all.

Despite the magical beauty and majesty of the African bush, it's often an all-consuming challenge just to function, let alone achieve anything, in Africa. Stultifying bureaucracy, incompetence, official obstruction, and low-level corruption slow any administrative business to the speed of cold molasses. Conservation is hardly ever high on a government's agenda, even among officials who have jobs in conservation. None of this is helped by the fact that Zimbabwe's economy has been in free-fall for most of the last two decades. The country's currency and banking system all but collapsed by the end of the 2000s. The government's fiscal policies created incredible hyperinflation that astounded even experienced economists. By the end of 2008, Zimbabwean money was losing half its value every twenty-four hours. Supermarkets routinely changed their prices several times a day in an effort to keep up with inflation. The Reserve Bank was issuing 100 trillion dollar notes, which became almost worthless as soon as they were printed. Small-denomination notes were commonly seen lying on the sidewalks in town; not even the street kids would bother to pick them up. Running a research project in this kind of environment became a challenge all of its own. Getting spare parts for our dilapidated research vehicles, which were pounded to pieces on the terrible corrugated roads, was nearly impossible. A tank of diesel cost 100,000 Zimbabwe dollars—if you were lucky enough to find a fuel station that had any fuel. We'd carry currency around in supermarket carrier bags and pay for things with huge wads of cash, just like in gangster movies. The distorted economy was terrible for wildlife conservation. There was a thriving black market for currency, fuel, food, and poached wildlife products. Anything that could be converted into hard foreign currency was too tempting for the unscrupulous or the desperate. Almost all the rhinos in Hwange were shot by poachers during this time—their horns cut off their faces and sold to illegal international wildlife traders.

In March 2000 the project suffered a tragic setback. I have painful memories of that morning. It was 8 a.m. and the merciless sun was already searing away the sparse cloud over Hwange. Sweat soaked my shirt as I jogged through the bush with other men who'd joined a search party. My heavy medical pack chafed at my shoulders. As we plowed through the scrub, the tall hibiscus plants shed irritating hairs that burrowed into our sweaty skin, their flowers standing out pale, sickly yellow against a powder-blue sky. Above us, a single-engine Cessna spotter plane circled. All of us were desperately looking for Andy Searle.

Word had come the previous evening that Andy and his helicopter were missing. He had made a routine flight to Sinamatella Camp the previous afternoon and had been due back home in the evening. The warden of Sinamatella, Norman English, reported over the radio that Andy had left on his return flight in the late afternoon, intending to track rhinos on the way home. It was less than an hour's flight, so he should have been back by sunset. Most likely he'd had a mechanical problem and had made an emergency landing. A few hours earlier, Trevor, who was copiloting the Cessna overhead, had told us he'd spotted Andy's helicopter.

"The chopper is down near Chebema Pan," he said, with a grim shake of his head.

We all hoped Andy was nearby, waiting to be rescued. But any optimism the search party harbored evaporated when we found the crash site. The helicopter was on its side, main rotor mangled. Andy seemed peaceful, blue eyes reflecting the equally blue sky, but the glint of his gold wedding band was a stark reminder of the devastated wife and son he'd left behind.

Andy's death shattered our small community. As a man, Andy was hugely popular. As a conservationist, he was passionate and professional. For a long time I did not realize how deeply I'd been affected by his death. But a few years later, when I got the news that another colleague had crashed his aircraft, I went into a state of shock and could not stop shaking. Andy was buried under a huge ebony tree at the edge of the park. His funeral was attended by everyone who lived and worked in Hwange. Conservation supporters flew in from as far as Australia. The African staff from Umtshibi sang a final praise song, a haunting melody that echoed across the empty bush.

I carried on the research work, as I'm certain Andy would have wanted. Our priority was learning more about the wider lion population, especially lions I had not already logged. The safari camps in and around the park were always a great source of information on the lions in their areas. One day, Dave Christensen, the professional guide who managed Makalolo Camp, excitedly told me about a new male lion that had been seen close to the camp. He invited me to stay for a few days and try to get a collar on him. He said the cat was shy and unused to vehicles so they had not been able to observe him closely. The coalition of males that had been resident in the area had both recently been trophy-hunted, and no adult males had been seen in the area for close to a year. I agreed to come to the camp in the next few days.

I'd recently been joined on the project by a university friend, Zeke Davidson. Since graduating, he had been working as an advertising executive in Cape Town and was burned out and needed a change of career. He came to spend a few months on the project to unwind; he ended up staying on and doing a doctoral degree in zoology. Dave used to call the two of us Dr. Jackal and Can't Hide—referring to my earlier jackal research and the fact that Zeke is a lanky six foot eight inches tall. We packed up our camping and darting gear and drove for three hours along the dusty park track. Makalolo Camp is a premium tourist lodge, with luxury tents set on platforms linked by raised boardwalks that thread through the edge of a teak forest, overlooking a wide-open plain sparsely dotted with mitswiri trees. The photographic concession, run by Wilderness Safaris, was bordered to the north by the railway line that marked the beginning of the hunting concessions.

The Makalolo area was once famous for lion sightings. Of late, though, lions had been scarce. Dave was worried about this, both because he was a dedicated conservationist and because tourists want to see lions more than any other species. As any safari guide will tell you, a game drive returning to camp without having seen a big cat is one carrying disappointed safari guests. Part of Dave's concern was that, over the last two years, three male coalitions had each been briefly resident in the exclusive 300-square-kilometer photographic concession. The males had moved in, mated with the resident pride, but in every case had wandered too close to the park boundary and been lured to their deaths by strategically placed trophy hunters' bait. The consequence of this high turnover of territorial males was that the prides had been unable to raise any cubs. Without the injection of new animals into the population, the prides grew smaller and more fragmented. The arrival of a new male gave Dave hope that he'd soon have a stable resident pride on the concession once more. When Zeke and I arrived in camp, he was virtually jigging with excitement.

"You're just in time," said Dave. "This morning Foxtrot saw the new male mating with a female from the pride. If we hurry, we could dart and get a collar onto him."

Fitting a radio collar to Johnnie Walker as Foxtrot looks on.

Foxtrot was a radio call sign for Foster Siyawareva, a tall, burly safari guide with an even bigger, booming laugh. A former soldier who had fought in the bitter civil war, he was completely at home in the bush and his sharp eyes missed nothing. Foxtrot soon confirmed Dave's account.

"Yes, I saw him this morning," he reported happily. "He was very busy with his new wife. Ho, ho, ho."

As there were only a few guests in camp, Dave and Foxtrot quickly volunteered to show us where the new male had been seen. When mating, male lions guard the female for several days, mating every ten minutes or so. In that time, they seldom move far. Their preoccupation with mating makes them easy to approach, though it is best to do so with caution. The testosterone-fueled males can become extremely aggressive. The opportunity was too good to be wasted. I quickly loaded up a tranquilizer dart. Half an hour later, the new male was wearing one of our new GPS collars. We gave Foxtrot the job of naming the new cat. After a moment's thought, he announced, "I will call him Johnnie Walker." Like the jaunty top-hatted fellow on the whiskey bottle, he explained, this lion "is always moving, always walking in search of new girlfriends."

A few days later, Zeke and I tracked Johnnie Walker to check up on him. He had lived up to his name, journeying far to the south of where we had caught him. It took several hours to find him by tracking the VHF radio signal transmitted by his new collar. He had followed the Linkwasha Valley that cuts through the flat, teak forest–covered landscape, deep into the Dzivanini wilderness and the protected heart of Hwange. Few people ever visit this part of the park, and it can only be accessed by a rutted, overgrown track. It is a remote land, empty of human presence. The only hint of humanity is your shadow reaching out across the sandy ground, a greedy invasion of the pristine wild. Elephant herds browse along the edges of the valley, as do small groups of zebras, tails swishing as they graze in the sun. Here the world is at peace with itself.

We found Johnnie Walker in the late afternoon as the sun dipped into the west, or at least we found the patch of scrub in which he had hidden for the day for shade and security. We did not disturb him further. As it was getting late, we decided we would drive down the valley toward Mandundumela Pan, the dry water hole at the far end of the valley, and camp for the night. As night fell, we pitched our tents under some stunted thorn trees and lit a campfire on which to cook. The night was warm and inky-dark before the moon rose. Above us, the stars of the clear Kalahari night were trapped in a spiny net of acacia thorns. As we waited for the smoke-blackened kettle on the coals to boil, we heard the bass rumble of a male lion's roar from up the valley. The cat continued to roar periodically, on each occasion getting closer to our campsite. We went to bed listening to his roars resonating along the valley. At midnight everything was once again quiet, but the wind was picking up and I woke to the rustling of grass outside my tent. I had a niggling worry that the coals of the campfire would flare up in the gusting wind and spread into the tinder-dry grass of the valley. Bush-fires spread rapidly through the dry bush and are almost impossible to extinguish, especially with a stiff wind behind them. I climbed out of my bedroll and stood up in the tent door in order to shine my bright Maglite flashlight around the camp. The fire was out, and I was about to return to my bed when I got a distinct sense that I was being watched. I felt a tingle between my shoulder blades, a warning reflex inherited from a primeval ancestor. I shone my flashlight around the camp and surrounding bush once more but could see nothing. After a few more scans, I crept back into my sleeping bag and was soon asleep again.

In the dewy early morning as we were breaking camp, Zeke suddenly exclaimed, "Look here! Lion tracks right in the camp!"

Sure enough, the unmistakable spoor of a large male skirted the edge of our camp and led to the far side of my tent. Judging from the tracks, the big male had stood there, a foot away from the flimsy nylon tent wall, before moving on. I have often wondered whether he was in fact there when I had sleepily checked the fire the previous night, blocked from my view by the tent. I am certain he was. The rustling of grass I had heard was in fact the approach of the big cat. This was a situation that could easily have gone badly wrong. Fortunately, the lion was not hunting and was only curious about the interlopers in his wilderness domain. It is also likely that the bright light made him wary and unwilling to investigate further. It was a thrill to have been so close to a big cat, even unwittingly—to feel a connection to a big, wild predator and be a part of its world. The experience also reminded me, though, of just how vulnerable an unarmed human is in the darkness that belongs to lions.

Johnnie Walker's career as a study animal was short-lived. When we remotely downloaded the location data from his GPS collar a few weeks later and plotted the locations onto a map, the spaghetti trail of hourly GPS points stopped abruptly just outside the national park boundary. There was no doubt that yet another study lion had been trophy-hunted. In something of a project record, Johnnie Walker had been collared a total of sixteen days before being shot. Dave and Foxtrot were crestfallen; their hopes that a male lion would establish himself in the Makalolo concession were once again dashed.

# FOUR

# KATAZA, THE TROUBLED ONE

I was devastated when Kataza was shot. It seemed a sorry end to a lion whose story I knew so intimately and who had survived against the odds for so long. Cats are supposed by some to have nine lives. Certainly cats of all sizes, from the domestic cat to the gigantic Siberian tiger, are tough, resilient creatures that have a way of looking out for themselves. Yet in the modern world, humans pose threats that wild cats are poorly adapted to surviving. Death at the hands of people is the most common and most significant cause of population decline among large predators, whether tigers in the Russian Far East, jaguars in the Brazilian Pantanal, or clouded leopards in deforested Borneo. They die because people shoot, trap, or poison them, killing for sport, pest control, or trade. People also decimate the prey that cats need to survive, or convert their habitats to farms and plantations.[19]

The story of Kataza, a young male lion that our project monitored for six years, epitomizes the threats lions now face. Local safari guides, who watched his tribulations, gave him his name, which translates roughly as "the one who faces troubles or misfortunes"—"*tadza*" being slang for an accident or mistake. Kataza survived many hazards—some natural, some at the hands of humans. Just as some people appear to

invite misfortune, Kataza seemed to tumble from disaster to calamity. Yet none of these perils was unique, not in a population where the lion's natural enemy was man.

When I first saw Kataza, in the second year of the project, he was one of two small cubs in the Dete Valley pride. The other cub was a female. We later named her Perdita, as her life was no less turbulent than Kataza's. The pride was regularly seen by tourists on game drives, and the safari guides would often find them resting up for the day in the stately grove of Camelthorn acacias at the end of the valley. The pride had three more cubs six months after Kataza and Perdita were born, with all five cubs sired by the Black Mane coalition. The coalition consorted with two prides and the unashamedly polygamous males would alternately visit their separate families in the Dete Valley and neighboring Safari Lodge prides.

Disaster struck the Dete Valley pride when Kataza and Perdita were a year old. Their young lives had been trouble-free until trophy hunters shot both Black Manes. The loss of the males made the pride

vulnerable to the attentions of new males, who would inevitably attempt to take control of the pride territory and claim the females as their own. The Black Mane coalition had, up until this point, roared and scent-marked assiduously in order to advertise that the territory was occupied. Without this auditory and olfactory billboard, any new territory-seeking males would know they could safely colonize the range without engaging in a costly and potentially dangerous fight. Ulaka, a large and aggressive six-year-old male, was at the time searching for a vacant territory. His arrival in the home range of the Dete Valley pride spelled disaster for the young lion cubs.

Ulaka, the cantankerous arriviste, caught the pride unawares early one morning. The females were away hunting and had left their cubs on their own at the edge of an acacia thicket. Ulaka ambushed them in a flurry of dust. Two quick bites accounted for two of the smaller cubs. The others scattered. Ulaka hunted them through the scrub, homing in on the last of the younger cubs that had made it into the thicket. He was relentless as he forced his way into the thorny bush. With Ulaka's focus on the cub in the thicket, young Kataza and his sister, Perdita, ran, and kept running. Once Ulaka was confident he'd killed all the small cubs, he emerged bloodied from the thicket, picked up one of the small, lifeless bodies, and carried it, almost tenderly, to the shade of a small blue bush. There he slumped, surveying the destruction he had wrought. He did not seem to notice the escape of the two older cubs. It had been a narrow escape for Kataza and Perdita.

The day after the massacre of the three youngest cubs, I found one of them in a pitiful state. He had not been killed by Ulaka, but the bite through his lower back had severed his spine, leaving him to drag himself painfully on his front legs. There was no chance he could be saved. Lionel put him out of his misery with one shot from his hunting rifle—a merciful end for an animal that would have died a lingering death if the hyenas or jackals had not found him first. In the aftermath of Ulaka's infanticidal attack, the two lionesses, along with Kataza and Perdita, vanished from the area for several months—apparently hiding from the killer male. Ulaka had by now turned his attention to the Safari Lodge pride, also left bereft by the loss of the Black Mane coalition. He soon demolished the ten young cubs in this pride too. With no young cubs to defend, the pride females soon mated with him and he set up residence in the pride's range—murderer turned family man.

Months later, with Ulaka's attention elsewhere, the Dete Valley pride slowly moved back into their old hunting grounds. The safari guides and I would occasionally also catch a glimpse of them. They were more wary and skittish than they once had been. Kataza and Perdita

Ukala consorting with a female from the Safari Lodge pride.

had grown long-limbed and lanky as they approached eighteen months of age. Kataza showed the fluffy signs of a sparse mane on his chest and an almost comical, Mohawk-style crest of hair on his head. They had not yet learned to hunt for themselves or to navigate the complex social system of their species. They were still completely dependent on the adult females to survive. Kataza in particular was vulnerable to attack. Adult males will kill even quite large subadult males that are not their offspring. The telltale evidence of Kataza's emerging mane marked him out as a future rival to be eliminated.

The two cubs, having survived the first deadly event of their lives, were about to be exposed to another. The area in which the pride had established its range was close to a large hotel, where staff lived in a nearby compound with their families. As is often the case in rural Africa, people supplement their household food supply with illicitly obtained bush meat. The killing of wild animals for subsistence goes back millennia, but modern methods of hunting for it are much more efficient. The most frequently employed method is to use nooses or snares made of galvanized steel fencing wire. These snares are set along the game trails used by wild animals, the noose is positioned over the path and the other end attached to a tree. Wire is abundant throughout Africa, harvested from fences or abandoned telephone lines. To gauge the extent of this problem, during several years we monitored how much wire had been removed from Hwange National Park's eastern boundary fence. We calculated that the 285 kilometers of galvanized steel wire stolen could have produced 71,000 snares.[20] Each of these snares, set to catch animals ranging in size from a giraffe to a small duiker, is capable of killing or mutilating many animals, since snares can be used many times. Abandoned snares remain, like low-tech land mines, in the bush for years. A successful poacher can make a good living selling bush meat, and in some parts of Africa, wildlife populations have been devastated by this kind of commercial poaching. Predators are equally susceptible to being caught in snares, and although their capture is usually unintentional, their body parts are readily peddled to those shopping for traditional medicines, amulets, and magical tokens.[21] Alongside the ubiquitous Kalashnikov AK-47 assault rifle, wire snares are the scourge of Africa's wildlife.

Tragically, the cubs' mother fell victim to a poacher's snare. With their mother dead, the cubs were entirely reliant on the remaining lioness in the pride. Their survival hinged on her ability to hunt enough food and keep them all out of the way of other lions or people. Several months after the death of the cubs' mother, a further calamity hit the unfortunate pride. Kataza, now two years old, was himself caught in a snare. From having seen sites where lions have died in snares, I can imagine what happened. As the noose tightened around his neck, the young lion leapt forward in fright—a move that only served to tighten the steel necklace. He fought against it, clawing at his neck

and the surrounding vegetation in his panic and terror. The bark of the tree to which the snare was attached was soon shredded with deep parallel scratches from Kataza's frenzied clawing. The implacable wire only became tighter with each frantic lunge, lacerating his neck with a deep wound into which the wire became embedded. With every panicked attempt he made to escape, the cat became weaker and weaker as the noose tightened. Then, perhaps because of a weak point in the wire or because it was poorly attached to the tree, Kataza made a last desperate wrench and was suddenly free. He bolted off through the bush, trailing a wire leash, the cruel noose still embedded in his neck but no longer tight enough to strangle him. Though he had escaped, the wire remained lodged in the terrible wound and prevented it from healing. Soon infection set in.

Around a week after Kataza's escape from the snare, Sibahle "Sibs" Sibanda, one of the safari guides, spotted the injured cat heading into some thick bush. Sibs called me on the radio. By the time I got to the spot he was last seen, the lion had disappeared. For the next two days, I joined Sibs and the other guides and camp managers in looking for tracks that might give us a clue as to where Kataza had gone. We found nothing. We could only wait for the lion to reappear. If he was still alive.

In the early morning, the third day after the injured cat had first been spotted, we found lion spoor on one of the dusty tracks through the bush. Sibs is an excellent tracker, and he soon interpreted the signs.

"There is a big lioness and two others," he said, examining the ground carefully. "See, here are the tracks of two females walking one behind the other. Another lion was walking behind them. See here. It is a young male."

Sibs's forehead creased in concentration. "Look, these scuff marks are made by the trailing wire," he surmised. "This is the snared lion. It is Kataza. We have found him."

The tracks were fresh, made perhaps twenty minutes before. I loaded up the dart gun so that we could immobilize the injured cat. We had to be ready; we might not get another chance. From the vehicle we followed the tracks along the sandy road for several hundred meters before they led into a thorny thicket we could not drive through. Entering the thick bush on foot to look for a badly injured lion would be foolish and could get us mauled or worse. We could only wait until he was in the open. Things started to look hopeless once more. If we did not catch Kataza, remove the snare, and administer antibiotics, he would certainly die.

Once again Sibs's legendary bush craft came to the fore.

"Don't worry," he said with an infectious grin. "I will call the lions here. Be ready with the dart gun."

Sibs cupped his hands to his mouth and produced a loud, drawn-out bellow, imitating exactly the distress call of a buffalo calf. Again and again he called, mimicking the rise and fall of a dying buffalo's terrified bawling. After a few minutes, we could discern the faint rustling of something moving in the bush and soon made out the form of a young lion, intently interested in the prospect of a free meal.

"Call again," I whispered to Sibs. "I can't dart him in the bush, he needs to be in the open."

Once again the sound of the dying buffalo enticed the cat toward us. Despite his injuries and the searing infection, the hungry cat could not resist the temptation of a free meal. He stepped unsteadily onto the road, providing me with the opportunity to dart him with the tranquilizer. Fifteen minutes later, once the drugs had taken effect, we surveyed the downed lion. The wound was horrendous. The snare had cut deeply into the muscles of the lion's neck, and the wound was an inch and a half deep in places. Sibs shook his head. There did not seem much hope of saving this lion without veterinary attention, and there was no hope of getting a vet to the middle of nowhere at short notice. We'd have to do the best we could. I felt into the wound and located the wire, which I snipped through with a pair of pliers from the toolbox. I pulled the wire loose. The gaping wound was pus-filled and infected, so closing it with stiches was not even an option, even if we had had the wherewithal to do so.

"I'll have to just clean it up and give him antibiotics," I muttered. "That's all we can do."

I packed the wound with iodine ointment and wound powder and injected a dose of antibiotics into his shoulder.

After the snare removal operation, Kataza disappeared. The two females were occasionally seen over the next week, but he was not with them. We started to believe that he had not survived the infection caused by the devastatingly deep wound. Just as I was thinking I would have to record Kataza as another mortality, an excited Sibs called over the radio.

"He is alive!" he proclaimed. "I have seen him with the others."

Sure enough, Kataza had reemerged. A furrow of short hair ran through his growing mane—the remnant of his snare wound. Yet, amazingly, he appeared to be otherwise unaffected.[22] Kataza had survived the third trial of his young life. But fate had more to throw at him.

The last remaining adult lioness in the pride disappeared, leaving the two half-grown lions alone. I was fairly certain she too was killed

Treating Kataza's snare wound.

in a snare. Some months later, a lioness skull was found in an area frequented by poachers, who by then had all but wiped out the pride. The two young lions, Kataza and Perdita, were just at the cusp of being old enough to survive on their own. However, without the adult lioness to hunt and provide food for them, they soon lost condition as they were far from efficient hunters themselves. Over the next few months I saw them often, and as necessity made them better hunters, they began to thrive. But without the support of experienced pride females, they could not defend a range and soon began to wander over large distances to avoid nearby prides and strange males. This wide-ranging movement soon brought them to the edge of the protected area and into contact with peasant farmers in the communal land that surrounds Hwange on two sides. Here agro-pastoralists keep small numbers of cows, donkeys, sheep, and goats, tilling small fields scratched out of the bush to feed their families. Life is hard for these people. Their livestock is often poorly protected, left to graze unattended close to the forest, or sometimes in the forest itself. Unprotected domestic animals are easy pickings for hungry, inexperienced predators, and the two young lions soon began killing cows and donkeys. Inevitably, angry farmers began complaining to Elias Mafu, the local National Parks warden, demanding action be taken to remove the lions.

Part of the role of National Parks is to protect people and their livelihoods from wild animals, and the usual response to livestock raiders is to shoot them. A team of rangers was duly dispatched to destroy the lions. Luckily for Kataza and Perdita, it was the middle of a particularly cold winter. When the rangers did not immediately find the lions, they returned home, having no appetite for spending several long, cold nights waiting for the lions to appear. At this point, I approached the warden and asked if I could catch the two lions and return them to Hwange National Park. He agreed to let me have a try, just as long as the lions did not cause any more problems in the interim. I recruited Ricardo Holdo, who was doing research on elephants in the area, and Paul, a lodge owner who helped me with fieldwork occasionally and was always up for an exciting lion capture mission. We were soon in action. The news came in the same afternoon that the lions had killed again near Magoli Village during the night.

The lions had killed the unfortunate donkey on the edge of a fallow field close to the village, but because of disturbance by people, they had moved into some thick scrub near a dry watercourse shortly after the kill. There was a chance, though, that they would return to the carcass, as they had not been able to feed much before being chased off. Of course, there was an equal chance they would move somewhere else and kill again. Lions that become livestock killers often do this, as they soon learn that to return to a kill is to risk being shot. I was banking on these lions being young and inexperienced and probably hungry, and therefore more likely to come back to kill. We waited in the vehicle, dart gun and spare darts at the ready, as the dipping sun streaked the horizon crimson. It was July, the middle of the Zimbabwean winter, and the temperature dropped as precipitously as the waning sun. Most people think of Africa as being hot and humid, but winter nights on the edge of the Kalahari can be bitingly cold. Recently, nighttime temperatures had dipped well below freezin and some nights had been so cold that small tree saplings were killed by the ground frost. Our chilly vigil dragged on for several hours, and still the lions had not come. However, I had a trick up my sleeve. I had brought a cassette player and a large speaker, along with a recording of a squealing pig. This is an awful screeching, keening noise that raises the hairs on one's neck and sets nerves jangling. Lions, though, are instantly attracted by this distress call and will often approach in the hope of snagging an easy meal. It worked. Just as we felt we could no longer tolerate the horrible sound, two gray-tawny shapes materialized out of the gloom. They were obviously hungry, and both quickly began to feed on what was left of the donkey. Kataza hardly flinched as my dart thumped into his rump. He fed a little longer before he began to nod, his head finally resting on the donkey carcass. Perdita was warier and kept to the far side of the carcass until, in a careless moment, she showed her shoulder and I was able to snap off a shot. Soon she too was tranquillized.

It was now past midnight. The air was chilly and frost was forming on the dry grass at the edge of the field. We loaded the two anesthetized lions into the truck that Paul had brought from his camp. We bundled them up carefully in blankets to prevent them from becoming hypothermic from the wind whipping through the open vehicle. It was perhaps a four- to five-hour drive to the place to which the warden had agreed we should move the lions. The release site he had authorized was in the middle of the park at Shumba Pan, 100 kilometers away. We hoped moving them far from their old home would deter them from returning, though it seemed likely they would not stay put. For this reason, both lions were fitted with radio collars so we could monitor

Transporting Kataza and Perdita.

their movements. We had a moment of unwanted excitement when we stopped on the way to the park at the deserted fuel station at Main Camp, the tourist hub of the park, in the small hours of the morning to check on the lions. As I was tucking the blankets around the cats, Perdita woke up and jumped out of the truck. Luckily, she was still groggy and we were able to dart her again and reload her. The drive to Shumba took the rest of the night, as we had to drive slowly on the badly potholed road, stopping frequently to check on the lions and inject them with additional tranquillizer. The dawn was painting the eastern sky in pastel reds when we reached our destination. We quickly unloaded the siblings in a clear area; then we sat a little way off in the vehicle, exhausted, sipping lukewarm coffee from a thermos. The lions slowly came around and, after seeming initially agitated, moved off into the safety of a Diospyros bush to sleep off the drugs. I doubted they would stay in this area. I knew there were other lions here that would soon see the strangers off. Possibly, they would move back toward their old home, driven by the deep-seated homing instinct all wild animals seem to have. I hoped, though, that they would not leave the park again. With limited resources and few other options, our intervention was not the perfect solution, but it had saved these two from being shot as live-stock killers. Within a few days, they did move back to their old home range, but this time they did not leave the park. Kataza had yet again narrowly escaped death, this time from the bullets of the National Parks problem animal control team.

We monitored the two young lions for more than a year. They sel-dom came close to the park boundary and appeared to have settled. Yet it seemed unlikely that, as siblings, they would stay together. It is the fate of young male lions to strike out on their own and find a pride and territory to defend. The day came when Kataza could not be found. Perdita was there as usual, but no amount of searching, even from my microlight aircraft, could detect a signal from his radio collar. Perhaps he had dispersed way beyond the area in which we operated, or perhaps he had left the park and been killed by a poacher or hunter. Poachers sometimes destroyed radio collars, or buried them, to silence their sig-nals, lest they lead game wardens to the scene of the slaughter. Some hunters even boasted that they had thrown collars from lions they'd shot onto the trains that rattled along the northern boundary of the park, taking the transmitters far beyond the reach of our radio receivers.

Two years later, Zeke Davidson, the project's doctoral student, and I were driving through the Shumba Pan area. It was late afternoon. We were rounding a bend in the road not far from where Kataza and Perdita had been left, when we came upon a male lion plodding down the road toward us. We did not recognize him as one of the males from the Shumba pride. The mystery intensified when, peering through my binoculars, I noticed that he was fitted with a radio collar. Who could this be? Zeke pulled out the radio receiver and scrolled through the frequencies of all the male lions fitted with collars that seemed likely to be in the area. Nothing. Suddenly something occurred to me. Could

Kataza at four and a half years old.

this be Kataza? At around four years old, the lion was the right age to be Kataza—if he had, against my expectations, survived.

"Try channel 40," I whispered to Zeke.

Sure enough a faint *blip, blip* came from the receiver speaker. The collar's transmitter battery was almost out of power, but at this close range it was just strong enough to emit a faintly detectable signal. Kataza, the survivor, had made it through the challenges and perils that a dispersing subadult male must face. I was thrilled to see my old friend again—a lion I'd twice rescued from certain death. Though I doubt he viewed me as his savior, I certainly felt an emotional attachment to him.

There was an urgent need to replace the collar, as the signal was weak and we would most likely never find him again if he kept moving. Being able to monitor his movements and behavior would give us a great deal of insight into the survival and fate of young adult male lions, especially since we already knew so much about this cat. Luckily we had the darting gear and a spare GPS collar with us. We soon had Kataza darted and fitted with the new collar. When I bolted on the new collar, I noticed he still had a bare patch around his neck where the snare had cut so deeply into his flesh.

Kataza set up a territory between Shumba and Nehimba Pans. As a singleton male, he could not hope to hold down a territory rich in prey. So he had to settle for a marginal dry area, with few females and scarce prey. Lion home ranges in this part of the park are huge because of the dispersed resources. Animals with large ranges are much more vulnerable than those with small, compact ranges, simply because it is much more likely that a large range will eventually extend beyond the sanctuary of the protected area.

Kataza's range drifted into the Deka Safari Area to the north of Hwange, an area managed for the purposes of trophy hunting. In safari areas, the National Parks authority manage the area and allocate a quota of wild animals that can be hunted each year to a professional safari-hunting operator. The operator then markets these to trophy hunters, most often wealthy North Americans or Europeans, who pay large sums to come on hunting safaris and take home "trophies"—most often the heads and skins. In theory, at least, this is a well-managed system in which small numbers of animals are legitimately hunted to generate large revenues that directly and indirectly benefit conservation. Indeed, hunting has formed the backbone of much of Africa's conservation strategy for decades. The Deka Safari Area had a one-lion quota for 2004, and a lion hunt was duly sold to a trophy-hunting client.

Kataza had no experience with hunters. To him, the zebra haunch hanging by a chain from a tree just over the park boundary was another free meal. A free meal that would be his last. On May 26, 2004, a trophy hunter shot Kataza dead. When we heard the news, we were all devastated. Of course we knew it might happen. All male lions living close to the park boundary are vulnerable, and we'd already lost several study animals to hunters.[23] But it seemed such a waste that Kataza could survive so much and yet still succumb to a hunter's bullet. To the hunter, no matter how exciting the hunt, this as just another lion. He would have no idea, as we did, of this animal's amazing life story. We felt robbed of an animal we had worked so hard to protect, in which we'd invested so much emotion and effort. Yet the killing of Kataza was entirely legal. It just did not seem right or fair.

# FIVE

# POACHERS, POLICY, AND POLITICS

After the first four years of research on lion behavior in Hwange, it was starting to become clear that many of the male lions we collared in the park were being shot in the surrounding hunting areas. These hunts were approved and completely legal. But it was puzzling and worrisome that so many lions, supposedly protected by the national park, were dying. I took my concerns to the National Parks ecologist at Hwange Main Camp. He was a pleasant, well-educated man who was supportive of our research. He was, however, to be found in one of the local bars as often as in his office at the park headquarters. The afternoon I came to speak to him, I discovered him at the Waterbuck's Head, the bar at the park's tourist rest camp. He had obviously been there a while, as evidenced by the row of brown quart beer bottles at his elbow. With the curtains drawn against the bright afternoon sunlight, the bar was dim and smelled of stale beer. The pre-sundowners lull was accentuated by the *thwop, thwop* of the ceiling fan high in the thatched roof. The ecologist peered at me through the gloom. His difficulty in focusing was not helped by the fact that he was wearing two pairs of glasses, one with the right lens missing, the other the left. He was having difficulty keeping them both in place with his forefinger while simultaneously maintaining his perch on the bar stool.

When I broached my concern about the hunting quotas, he appeared annoyed. Of course the quotas were not too large, he scoffed. They had been approved at stakeholder quota-setting meetings with the landowners and signed off by the minister of environment himself. In fact, he went on, the same quotas had been given for years, and nobody had ever questioned them. If I did not believe him, he yelled irritably, I should go and check the records kept in a file in the park's administrative office.

Thus authorized, I did so. What I found alarmed me. Leafing through the files, I could see that each of the thirty private properties in the Gwaai Conservancy bordering the park had applied for a hunting quota.[24] Added to this, quotas were also allocated to the state safari-hunting areas, to forestry land, and to two adjoining rural district councils. In total, thirty-seven relatively small parcels of land had hunting privileges. These "quotas" were supposed to be based on an evaluation of wildlife populations resident on the property or the concession in question. Invariably, it seemed, each landowner asked for and was allocated at least one male and one female lion to hunt. For larger properties, between two and five male lions were allocated.[25] Given what we now knew about the ranging behavior of lions, it seemed improbable that each and every relatively small property had its own resident lions. It was much more likely that a few lions were roaming across a wide area encompassing many small, unfenced tracts of land. If indeed quotas were based on any credible population estimate, it seemed that landowners were double, triple, or quadruple counting the lions moving through the area. We already knew that most lions that were shot, at least the adult males, were animals that normally dwelled in the national park and only made short forays into the hunting areas. We knew this because we'd fitted most of the lions being hunted with radio collars and we could see from the data that they spent most of their time in the national park. It seemed hunters were counting the same lions on their land that we were studying in the park. All in all, sixty-two male and thirteen female lions were on quota to be hunted in 2003 in the hunting concessions around Hwange; seventy-five lions were eligible for killing. This was marginally lower than in 2002, when seventy-three males and thirty-six females—129 lions—were allocated to be hunted.

This was astounding. It seemed nobody had ever thought to add up all the lion-hunting quotas to see what kind of impact hunting might be having on the overall population. Based on male lion home range sizes and density of males in our study area, rough estimates suggested there could be no more than thirty adult males in the entire national park population. In effect, every single one was on the hunting quota and could be shot as soon as it poked its nose over the park boundary. This chimed well with what we knew about the survival of male lions we had radio-collared. I calculated that, on average, male lions had a 50 percent chance of survival in any one year, though in some years the survival rate was as low as 27 percent.[26] These animals were not dying of old age, disease, injury by dangerous prey, or in fights with other males, as one would expect in a naturally functioning national park ecosystem. They were invariably being shot by hunters outside the park. It also made very little difference whether they were adults or young males. If it was male and had half a chance of being passed off to a wealthy hunting client as a trophy, it was "fair" game. Of course there was no prospect that the annual quota could ever be filled; there just were not enough lions. In all, around sixteen male lions were shot each year, or more than half the adult males in the population. This did not seem sustainable.

Armed with my findings, I once again met with the park ecologist, this time in his office. No mention was made of our less formal meeting in the bar several days before. As he scanned the figures I had compiled, his brow creased with concern.

"Yes," he grunted. "This is a problem."

To his credit, he suggested that we should present this to senior officials at the National Parks head office. A few weeks later he called me to say we were to go to Harare and present the findings of the lion research project to the director general of Zimbabwe's National Parks Authority.

The National Parks Authority headquarters in Harare was a well-appointed building nestled in a spacious plot dotted with indigenous Msasa trees. It was next door to the city's beautifully manicured botanical gardens, not far from where I grew up. The plush administrative offices had been built by the European Union in the early 1990s, a step up from the drab prefab compound near the city center that had formerly housed the headquarters offices. In September 2004, on the day of our meeting, the new buildings were already showing signs of dilapidation and neglect. We stood in the large lecture theatre in the main HQ building, in front of a surprisingly large group of senior national parks managers and officials, nervously wondering what kind of reaction we'd get to the presentation we'd put together. The director general, Dr. Morris Mtsambiwa, and George Pangeti, chairman of the board, sat in the front row. After a brief introduction from the chief ecologist, I detailed how most of the lions we collared in the park were shot on the other side of the park boundary; how quotas were exceedingly high in relation to the number of male lions in the population; and, as a consequence, how survival rates of adult males were worryingly low. Dr. Mtsambiwa stood up at the end of the talk. He was an astute, bespectacled man with an affable, easy manner. His background was in fisheries ecology, so I could see that he had followed the presentation and understood the implications of overharvesting a resource. He thanked me warmly for the work the project had done to date, its value to wildlife conservation, and expressed his concern about the hunting situation. He promised something would be done.

A few weeks later, back in Hwange, I was called for a meeting with the park ecologist. With a big smile, he announced that, based on the evidence we had presented, and with the approval of the parks board, the director had suspended lion trophy hunting in the hunting areas around Hwange National Park for four years.[27] Not only that, lion hunting was to be suspended across the entire province in which the park was situated, Matabeleland North. He requested that we continue to monitor the population and report on the biological response of the population to lower levels of hunting mortality.

Members of the project were delighted and astonished in equal measure. This was a gratifying recognition of the work we'd done and also a wonderful opportunity to monitor the effect of reduced hunting on the lion population. I had not expected such decisive and responsible action. I'd thought there might be a reduction in the lion quota, given that it was clearly grossly unsustainable. But because significant revenue was derived from lion hunting, I suspected it would take several years for any change to work its way through the system. Now without such high levels of mortality, it seemed highly likely that the lion population would rebound and this would be exciting and scientifically informative to document. The guiding principle of wildlife conservation in Zimbabwe has always been "adaptive management"— that is, modifying management actions in response to monitoring data and scientific findings, and in turn monitoring the outcomes of these adaptations. This is exactly what was happening here. Scientific evidence: high quotas had led to over-hunting of lions. Management action: temporarily suspending hunting; monitoring the effect of new management policies. Outcome: use of data collected to implement a more appropriate management policy. I was happy to be contributing to a functioning wildlife-management system that was benefiting the species and conservation management in the country.

Unfortunately, my sense of satisfaction was short-lived. If I had been more experienced, less focused on ecology, and more alert to the politics of conservation, both national and international, I may have been more skeptical. Conservation is inextricably embedded in a convoluted mire of intrigue. Seemingly straightforward solutions that manifestly benefit wildlife on the ground are not always welcomed in situations where there are strong vested interests, both financial and political. The next few years would be an education in how conservation really works; where there is money at stake and corruption is rife; where deals are done between lobbyists and politicians behind closed doors at international conventions; where opposing worldviews collide, driven by intractable, financially entrenched lobby groups; where balanced scientific debate is replaced by disrespectful, vitriolic clamor. All this to the ultimate detriment of the wild animals everyone is purporting to protect.

The international backdrop to what was happening in Hwange was an augury of what was to come. In the run-up to the thirteenth international Conference of the Parties of the Convention on International Trade in Endangered Species of Wild Fauna and Flora (CITES), Kenya, a country that had banned all wildlife trophy hunting in 1977, tabled a motion to move lions from Appendix II (which allows regulated trade in a species between the 183 countries that are CITES signatories) to Appendix I (which in all but exceptional circumstances prohibits trade). This caused a flutter of apprehension across the trophy-hunting fraternity. A big motivation for most trophy hunters is shipping parts of the hunted animal (the "trophy") back home. Import of lion trophies might be prohibited in some countries if the species was placed on the Appendix I listing. This might deter many would-be lion trophy hunters and thereby have a big financial impact on the hunting industry. Although trade in Appendix I–listed species is generally prohibited, export and import, usually against stringent quotas, is possible with agreement from other signatories to the convention. Sport-hunted leopards, African elephants, black and white rhinoceros—all Appendix I–listed can be exported in very limited numbers from countries deemed to have secure, well-managed populations. Nevertheless, hunters were worried that if lions were elevated to Appendix I, and against a background of concern about the status of lions in many African countries, severe limitations would be placed on lion hunting over much of the species's range. Many conservationists in Africa sided with the hunters. Their concern was that if lions lost their value as hunting trophies, the incentive to protect lion populations across much of their range would be lost. As a result, lions might be deemed to be pests or vermin, as they had been in the past, and many populations outside strictly protected national parks might be eliminated. On the other side of the debate were those who felt strongly that lions and lion populations were in enough trouble already without rich people coming to shoot them. They were joined by animal-welfare lobbyists, widely believed to be behind the Kenyan proposal in the first place, who viewed recreational hunting in any form as deeply repugnant. The battle lines were drawn: on one side a few African countries, backed by animal-welfare groups, who favored better protection from uncontrolled trade in African lions; on the other, hunting lobbyists and others who believed lions were better protected if they retained a value to local African economies through the trophy-hunting industry. The debate quickly became acrimonious. Ultimately the tabled amendment that would see lions join Appendix I with endangered species in need of protection from harmful trade failed to gain support. However, the argument left lion conservationists divided and African wildlife managers from countries where hunting was part of wildlife-management policy deeply suspicious of the anti-hunting lobby.

With the Hwange Lion Research Project growing fast, there was a need to find funds to support the fieldwork. We needed to continue the research for at least the next four years to monitor the effects of suspended trophy hunting. Although the director of the Wildlife Conservation Research Unit, David Macdonald, was a fund-raising magician and a hugely supportive partner in the lion project, funding field-conservation projects was never an easy task. We'd often run the research on a shoe-string budget that barely covered the field staff's meager salaries and kept our old, creaking research vehicles out of the local mechanic's workshop. Several people suggested we should approach the Safari Club International (SCI) for support. SCI is an organization with a 50,000-strong membership of hunting enthusiasts that purports to support conservation through trophy hunting. Their publicity material boasted of their commitment to wildlife protection and their conservation successes, many of which were credible and genuine. Lions are of symbolic importance to international hunters, and SCI's charitable foundation's logo is a lion against a backdrop of an African war shield. Given that we were working toward more sustainable management of lion populations, surely SCI would be excited by our lion conservation work. Zeke Davidson and I were eventually put in touch with a gentleman by the name of John J. Jackson III, onetime president of SCI and now president of hunter advocacy group Conservation Force. He invited us to meet him at the SCI Convention to be held in January 2005 in Reno, Nevada. He'd arrange for us to address a forum of hunters and present our conservation work.

Zeke and I duly flew to Reno. We had a breakfast meeting with John at one of the big casino hotels. John is a slight, white-haired gentleman, a lawyer by trade and a vigorous advocate for hunters' rights. Over an early morning breakfast, we told him about the research and how we'd worked with Zimbabwe National Parks Authority to temporarily suspend trophy hunting in the area around Hwange National Park as a measure to allow the lion population to recover. John started to look a bit queasy, and not because of the greasy breakfast. A conversation about suspended trophy hunting was not what he had expected, and this was clearly not welcome news. Things went downhill from there. We were due to speak to a group of professional hunters and safari operators, mostly from Zimbabwe. They listened politely to the presentation I had prepared, modeled closely on what I had presented at National Parks headquarters at the end of the previous year. When I finished, there was a stony silence, followed eventually by a few skeptical questions. At the end of the meeting, one of the hunters summed up the mood with a comment I will always remember. He said, "Well, maybe hunting does have an effect on lion populations, but we can't let the anti-hunters see any of these results." We were left with the sense that they thought we

Presenting our research at the SCI Convention, 2005.

were pretty cheeky coming to the SCI Convention, the bastion of all things hunting, to criticize how hunting was managed. Our impression was that inconvenient scientific data was not going to get in the way of business. And business was what they were there for.

Later in the day at the SCI Convention at the 500,000-square-foot Reno convention center, we could see why. The four-day convention is a huge marketplace, and the three halls, each the size of a vast aircraft hangar, were filled to capacity with stalls selling hunting safaris. For sale were expeditions to hunt polar bears, buffalo, lions, elephants, exotic sheep—pretty much any species it is possible to hunt. Intermingled were stalls selling expensive, bespoke hunting rifles, expedition and hunting gear, wildlife art, trinkets, and taxidermy services. Many of the hunting company's booths were decked out as replicas of their hunting bush camps, camp tables arranged with glossy brochures featuring the previous season's hunts, and offerings and price lists for the coming

On the floor at the SCI Convention, 2005.

hunting season. The crowded lanes between the stalls were thronged with hunters looking for the best deal on their next hunt. They haggled with hunting operators and outfitters and chewed the fat with old friends. In amongst the taxidermied hunting trophies were the trophy wives, dripping in expensive jewelry, tottering after their aging Actaeons. The wealth on display was astonishing, especially for two lowly biologists straight out of Africa. This was a place where animals, at least the lives of animals, were bought and sold on a colossal scale.

We'd arranged to display a poster about the research project at a stall run by the Safari Operators Association of Zimbabwe, the body that represents photographic and hunting tourism companies in the country. I recorded in an email to Joanne that "we [spent] most of the time standing at our booth and answering questions about Zim. A few people [were] interested in the research [but] not many." We'd been naive to expect that the hunting fraternity would be falling over itself to support research that was exposing some of the flaws and excesses of the industry. This was not something hunters wanted to know about. Most certainly, in light of the CITES debacle the previous year, they did not want overhunting of lions to become public knowledge. There were other conservationists at the convention, mostly biologists funded by SCI, and several were attending a meeting on lion conservation organized by SCI. We were conspicuously not invited to attend. News of four years of suspended lion hunting in western Zimbabwe had alarmed the organizers greatly, and we were being kept very much at arm's length. To many in the hunting industry, there are two kinds of people: those who hunt and those who do not. Anyone who expresses an opinion that does not support hunting, is termed "an Anti." This is short for anti-hunter but could equally mean antichrist for all the venom spewed at critics of hunting. Our perspective on the management of lion conservation aimed to be entirely objective; we were collecting data to support wild-life-management decision-making. But the fact that our research had precipitated a hunting ban, even a temporary one, put us very much in the Anti camp.

Craig Packer, the eminent lion biologist who had studied lion biology for much of his distinguished career, was being feted by the czars of the SCI hunting world. He was launching Savannas Forever. This was to be an independent NGO that would certify and accredit trophy-hunting companies that could demonstrate they operated according to strict ethical guidelines aimed at managing hunting resources sustainably. The NGO would operate along the same lines as certification of wood and forestry products by the Forest Stewardship Council, a global certification system that allows consumers to verify that products they buy come from sustainably managed forests. His plans were impressive, and it was obvious that successful accreditation of trophy hunting would go a long way to eliminating some of the less desirable behavior in the hunting industry. Craig had heard we were there and dropped by our display for a chat. As always, he was intellectually hyperactive and full of energy. He was excited that his ideas on how to manage hunting sustainably seemed to be gaining traction. It later turned out the hunting industry was less enthusiastic about the level of independent scrutiny Savannas Forever proposed. As Craig relates in his book *Lions in the Balance: Man-Eaters, Manes, and Men with Guns*,[28] the vested interests that ran trophy hunting in Tanzania, where the program was to be implemented, quickly undermined the program. Craig lost his research permits and had to leave his study site in the Serengeti and the lions he'd dedicated much of his professional life to understanding.

The Convention was a dead loss in terms of raising funds. We left Reno without a single cent to support the research project. We didn't even cover the costs of being there. But being at the heart of the hunting world was an experience that allowed us to understand the hunting industry from a different viewpoint. As much as conservation is promoted, the industry is really about business and money, and the buying and selling of animals. It is a hugely lucrative industry that believes its own propaganda on the primacy of hunting in conservation. It does not want to change.

With a temporary ban on hunting, we went from a situation in Hwange in which any male lion leaving the national park was in danger of being indiscriminately shot to one in which adult lions were relatively safe. Because male lions were now living much longer, with the survival rate of males increasing to more than 80 percent, there were many more males in the population. More males meant lion coalitions divided the available space into smaller territories. In our core study site of nearly 3,000 square kilometers, we now had seven male territories, where there had once been only two. Instead of a coalition of males gaining tenure of several prides of females, each male group consorted with a single pride. The structure of the population was starting to look much more like those seen in well-protected national parks like the Serengeti or Kruger.[29]

Taxidermied trophies on display at the SCI Convention, 2005.

At the same time as we were starting to monitor the recovery of the lion population, Zimbabwe was going through an intense period of change. Over the previous few years, generations-old issues of land ownership had come to a head. Politicians, to cement their prospects of retaining power, encouraged the forcible eviction of white landowners. The land was reallocated, in theory to dispossessed, black small-scale farmers. In practice, it went largely to political heavyweights. The chaos of the so-called land invasions came to the Gwaai Conservancy in 2000 and all the former white owners were dispossessed, many violently. Skwatula and his family were also evicted. I was sorry to see him go. Although he was a hard man, I had become fond of him, and he was passionate about the wildlife on his farm. The lion project itself was evicted from the small farmhouse we rented on the border of the park—by none other than the provincial governor, who claimed the house and the land it was on as his own. The new owners of the farms were city politicians, generals, and military officers and their relations. They had little experience in managing the lucrative wildlife populations that had been nurtured on the ranches and farms in the Gwaai Conservancy. Poaching, previously well controlled by the former owners, became rampant. The chaos attracted the mavericks and opportunists who follow closely on the heels of social and political upheaval. The Gwaai Conservancy had a carefully cultivated reputation as a wildlife area and destination for international tourism. This was not lost on unscrupulous hunting operators out to make a quick profit. They quickly closed in to make deals with the new landowners. One of these was the notorious, whimsically named South African hunting company Out of Africa Adventurous Safaris. The owner was a South African ex-policeman–turned–professional hunter. The company and its directors had been implicated in rhino poaching in South Africa and prosecuted for smuggling illegal hunting trophies into the USA. This was one of a number of South African companies marketing trophy hunts in the concessions around Hwange to foreign trophy hunters.[30]

Although lion trophy hunting was officially suspended in the concessions surrounding Hwange, inconvenient restrictions such as this meant little to unprincipled hunting companies and their politically connected partners. Rules could easily be circumvented. Officials, paid off for the necessary permits, turned a blind eye to any hunting excesses. It was only a matter of time before illegal lion hunting started to happen.[31]

Late in 2005, despite the moratorium on lion hunting, a trophy hunter shot a collared lion we called Vuka. Jane and I had collared him deep in the core of Hwange at Mandundumela Pan in 2002. "*Vuka*" means "to wake up," as he had a habit of charging tourist safari vehicles, compelling guides to step on the gas. His range now encompassed an area outside the park in the Gwaai Conservancy. The radio signal on Vuka's GPS collar was no longer functioning, and for several months, we had rarely been able to locate him. Then he simply disappeared, which seemed odd given he was a territorial male. We did not at the time suspect he had been hunted. After all, there was a hunting ban in place.

I was later anonymously provided with a video of an American hunter's safari trip. Professionally made videos recording trophy hunters' safaris are a lucrative side industry. This video—essentially an animal snuff film—does not make easy watching. In one scene, to the accompaniment of an upbeat African soundtrack, the American shoots a running warthog from the back of the hunting vehicle. He does not make a clean kill; the bullet from his high-powered rifle hits the warthog in the abdomen. The stricken creature stumbles on, tripping over its intestines before collapsing in agony, still very much alive. The hunter callously drawls that it is a "bad day for the pigs." The videographer, I'll call him Video Man, records the dispatch of several other species, including a hippopotamus shot in a very small dam—the hunting equivalent of shooting fish in a barrel. The video builds to the crescendo of the hunting safari. The white South African hunting guide is seen organizing his black assistants to hang a reeking antelope carcass in a tree. This is bait to attract a large predator; from the height it is suspended—about two meters off the ground—it is clear they're gunning for a lion. Cut to a night scene. The American is installed on the back of a hunting vehicle with a view of the bait thirty meters away. Seated next to him is the South African guide. A male lion roars nearby and the occupants of the vehicle tense up. Soon a lion with a distinctive black mane can be seen approaching the bait. He ignores the vehicle and people and starts to feed. The hunter readies himself, waiting for the word from his guide to take the shot.

Video Man is startled at the deafening report of the gunshot; his camera jerks upward, then sideways, and we lose our view of the lion for a moment. Composing himself, Video Man focuses his wayward camera on the now stricken animal. The lion is lying on his side, legs thrashing, biting desperately at a clump of grass as he tries to comprehend the painful hammer blow he has received. The strength ebbs from his powerful body. The white hunting guide quietly reassures his excited client: "He's going down. Ja. He's done."

The South African is calm and authoritative. He's seen this before. Awkwardly, the lion lurches to his feet in a last attempt at escape.

"Hit him again," the guide orders.

Another shattering report. The video camera again shakes. When Video Man reframes his shot, we see a recumbent lion, now completely still. Poignant music seeps in. Rifle at the ready, the hunting guide and his assistant cautiously approach the downed cat. Video Man is in tow, assiduously recording the moment for the American waiting in the vehicle. He films the lifeless cat, its chest and neck slick with rapidly congealing blood. Despite the blood on its mane, we see a GPS collar. The hunter gestures at it with his rifle, seemingly puzzled. The collar has miraculously disappeared when Video Man frames the shot of the heavy carcass being heaved into the back of the hunting pickup by three men. Clearly visible: the gray tag I'd punched into the lion's left ear three years earlier at Mandundumela.

The film cuts to the lodge where the hunting party is staying. The dead lion is tastefully laid out on the well-watered lawn, his head resting on its paws, as if asleep. The worst of the gore has been rinsed from his dun-colored hide. The lodge's electric lights cast stark shadows across the hunter's face as he examines his trophy. Lodge workers dance enthusiastically around a blazing campfire, singing in praise of the

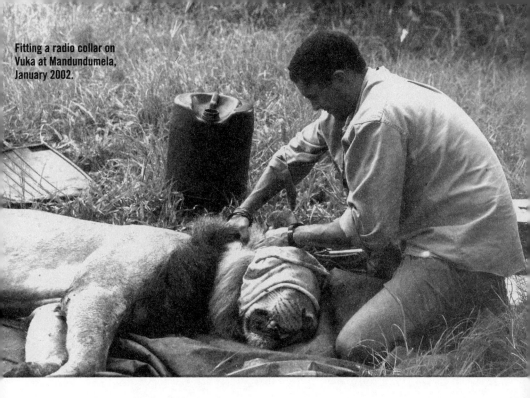

Fitting a radio collar on Vuka at Mandundumela, January 2002.

returning hunters. Perhaps they are genuinely pleased that this lion has been killed. Perhaps they have an eye on a decent tip from this wealthy client. The American grins foolishly, seeming oddly uncomfortable.

While we were completely unaware of this hunt at the time, it was to have significant implications for our project. The GPS collar vanished and we never saw it again. There appeared to be no official record of the hunt logged at any of the National Parks offices. It was even rumored that a high-ranking government official was behind the whole thing. However, the fact that the lion was collared, and potentially being monitored, obviously made the South African hunters, their local partners, and their political backers very nervous. They had disposed of the GPS collar, but they had no way of knowing it was not working, and they did not know how much we knew.

One of the outcomes of the CITES meeting was the resolution that lion populations across Africa should be thoroughly surveyed and enumerated. This led to a series of lion range–wide priority-setting meetings, jointly funded by SCI and International Union for Conservation of Nature (IUCN) and run by Wildlife Conservation Society biologist Luke Hunter and hosted by John Jackson from Conservation Force. These meetings brought together lion experts and conservation biologists, each armed with information about lion populations in the areas they worked. I was invited to attend the southern African component of the meeting held in January 2006 in Johannesburg, where I worked with other biologists to compile everything we knew about lion populations. The treasure trove of information collected at these meetings was all duly mapped and collated. It showed for the first time what many had suspected: lion populations were much more fragmented than previous distribution maps had shown. They had disappeared from much of their former range on the continent. In areas where little research or monitoring took place, their population status was extremely uncertain. In others, particularly in West Africa, lions were clearly imperiled.

The Johannesburg meeting was followed by a conference, held under the auspices of the IUCN, and attended by African conservation and environmental officials and government representatives. I coauthored a conference paper on trophy hunting with Craig Packer and several other biologists, to be delivered at the meeting to provide background to the officials.[32] We included an analysis of the levels of hunting in each country from CITES records, compared with estimates for each country's lion population. Most countries permitted hunting of a small number of lions, mostly less than 2 to 3 percent of their national populations per year. However, records indicated that Zimbabwe allowed a far higher proportion of its lions to be hunted annually. Given what we knew about hunting quotas around Hwange, this was not surprising. It was recommended that hunting should be cut back. I learned later that the Zimbabwean government officials were furious at having been singled out and embarrassed in front of their colleagues from other African countries. In retrospect, we should have been more careful about how this was presented. However, at the time, given how receptive Zimbabwe National Parks had been to acting on scientific data, I did not think this would be problematic.

Any seasoned political operator would not be surprised by what happened next. But for a bunch of conservation biologists, it was a hammer blow. A month after the Johannesburg conference, we got a phone call from the National Parks' chief ecologist in Harare. Our research permit had been revoked. We were to stop our research work and leave Hwange National Park immediately. When asked for the reasons behind this, the chief ecologist became agitated and told us he was unable to give us this information. He intimated that the orders had come from high up. We asked what we could do to have our permit reinstated. He told us there was no chance of this happening. The decision was final.

After all we had invested in the project, we were completely devastated. We were to abandon the study and leave the lions we had come to know so well—the lions whose lives we had come to understand so intimately. Zeke was in despair. He was in the middle of the field-data collection for his doctoral thesis, and losing the permit meant he could not finish his studies. Nevertheless, we packed up and left the park. We had no alternative. However, we did not give up the hope that we could reverse the situation. A few months later, we met the chief ecologist face-to-face in his office—a meeting he'd managed to avoid up until that point. He repeated his previous statement that we would never get the permit back, but he seemed to soften his stance slightly, which gave us some hope.

Meanwhile I arranged a meeting with George Pangeti, the chairman of the National Parks board, the governing body that oversees the running of the organization. He runs a small private school in the suburbs of Harare, and he suggested that we meet at his office there. This meeting was a lot less daunting than I had expected. Pangeti was courteous and friendly, and after listening to what I had to say, he acknowledged that he was aware that there might be some concerns in government and in the hunting industry about the research we had been doing. He would, however, look into this personally and let me know if anything could be done. When I called the next day, he told me that it seemed likely that the issues could be resolved and we would get the permit back.

He suggested I meet with the director of conservation at National Parks headquarters. The director had been one of my father's students at the University of Zimbabwe. I went to his office. On his desk was a folder full of papers. Papers that I could see included our old research permit as well as a sheaf of printed emails. Clearly there had been considerable correspondence. He thoughtfully fingered a printout with the name of a prominent international hunting advocate on the header. He said concerns had been raised at the recent lion-conservation workshops in Johannesburg about the purpose of our research on lions. However, my recent meeting with the chairman had clarified this and National Parks was now prepared to allow us to continue with the project in Hwange. We never did find out exactly what the problem had been and why this could not have been resolved without withdrawing our permissions. It was obvious, though, that pressure had been placed on National Parks to revoke the permit. We did not know whether this was at the instigation of politicians involved in the illegal hunt, because of unwelcome exposure of Zimbabwe's hunting policies or because international hunting interests viewed our research as a threat to the continuation of lion hunting.

# SIX

# PEOPLE OF THE LONG SHIELDS

Nkone cattle, the sturdy domestic stock of rural southern Africa, are a kaleidoscope of colors. Warm red-brown or white spangled with black, or white and red, or plain glossy black. They are long-horned and wild-eyed. A Ndebele herder knows every animal in his care, and sometimes gives them whimsical names to describe their coloration and patterning. *Umlotha* is the all-white beast, named for the bright white ash in the fire pit on a cold winter's morning. *Buthungu* has a white face. *Inhlavukazi* is the all-brown cow with the upward-pointing horns. The one with small black spots, like those on the lithe genet cat, is *Inkoni*. *Iqamuli* is the brindled cow, named for its likeness to the speckled pattern on the throat of the deadly spitting cobra.

In these herds lie the prestige, wealth, and reputation of their owners. They are a kind of African savings account that advertises wealth and status, to be dipped into only when necessary—say, to pay *lobola*, or bride price, when a son or nephew is married. They are also used as draft power to pull a plow or cart. Without draft animals, a family must till the field with hand hoes, backbreaking work in the October heat. When the rains come, a family who cannot plow cannot plant a crop of maize or small grain, and their children will be hungry.

Nkone cattle.

Cattle are part of the fabric of rural life. The secure enclosures where they sleep—called *bomas* or *dangas*—are built next to the family living quarters so livestock can be protected at night from predators and theft. The dull, clanking *tonk, tonk, tonk* of the bells that swing from the cows' necks as they move to their grazing pastures is the peaceful sound of pastoral Africa. In the evening, herders use the sound of the bells to locate the grazing cattle and bring them safely home. Not every cow has a bell; a few per herd is enough to round up the animals at the end of the afternoon. Sometimes a particularly wayward beast, with a reputation for being ornery, also gets a bell so it can be located more easily when it strays.

Farmers who live on the edge of the wilderness, next to national parks, share the landscape with others whose ears are as attuned as their own to the sound of a cowbell. Lions, spotted hyenas, and leopards are opportunists. Noises and smells are clues that guide their daily hunt for food. For a decade, the research project has collected reports of predation of livestock on village lands around the park. When I started to analyze the data, it was immediately obvious that a high proportion of cattle killed by lions and hyenas had been fitted with bells.[33] It seemed that predators had learned to use the sound of the bells to seek out the unfortunate cows—the African equivalent to Pavlov's dinner bell. If cows bought life insurance, cows with bells would be paying a much higher premium than their more silent herd mates. In fact, lions with livestock-killing proclivities are so attuned to the clank of a cowbell that it is quite possible to use the sound to attract their attention. We've sometimes used this trick to call in livestock killers we needed to fit with telemetry collars.

Conflicts between livestock owners and predators are as old as the first domestication of animals 10,000 years ago in the Fertile Crescent on the banks of the Euphrates. The turning point in humankind's evolution was when people tamed animals for the first time and vowed to protect them as their own. The conflict is as bitter now as it was then. Just ask any rancher in the American Midwest when a wolf pack kills a prize calf, or the sheep farmer in Scotland who has lost lambs to a hungry fox. Livestock owners the world over do not love predators. Retribution for losses is frequently swift. It is just that nowadays people have better tools to exterminate predators than did our Bronze Age ancestors. The conflict only stops when there are no more predators. For this reason, large carnivores have widely been exterminated wherever people farm livestock and have the means to rid themselves of species that are, in their eyes, vermin. This is one of the huge challenges that conservationists face in the few places where large predators and livestock owners still coexist. Livestock killing by lions is a frequent occurence in the farmland surrounding Hwange National Park. As part of the lion project's conservation work, field staff investigate and log these events.

The remains of the two black steers lay side by side, hollowed-out rib cages bared. Most of the flesh had been efficiently gnawed away by a small pride of lions. Only the long-horned, glassy-eyed heads were still intact. The scent of fresh blood and the tangy-sharp odor of gut contents were heavy on the breeze. The unmistakable pug marks of the big cats still showed in the sand around the carcasses, leading back toward the national park boundary fence, marked by the line of acacia trees in the middle distance. The owner of the dead cows looked on forlornly. He'd realized last night that the cows had not come back to the homestead. He'd come out at first light to find them, fearing and discovering the worst. The lions had killed the animals the evening before and fed on them all night; there was little left to be salvaged from the carcasses. The ensuing conversation followed a familiar pattern.

"That one," the owner of the cows says sadly, "she was with calf."

He wears the haunted look of a man who has been robbed, the hope of a small increase to his herd torn away. A small crowd of villagers soon forms around the dead beasts, scrawny village dogs sniffing around for scraps the scavengers have missed. The old man with one leg, Mr. Nkomo, the schoolteacher's father, starts to ask some pointed

Cattle killed by lions in village lands bordering Hwange.

questions.[34] Questions he already knows the answer to, but he needs to make his point and maintain his reputation as the village curmudgeon.

"You are from Lion Research," he states firmly, pointing at Jane and me with a gnarled forefinger. "Why do you allow Jane's lions to leave the park and eat our cows?"

He is convinced that because Jane has caught some of the lions in the area and collared them, they now belong to her. How else would they tolerate a collar? We have had this discussion before with Mr. Nkomo, and he understands what we do. But as the village elder, it is his right to maintain the pretense for the benefit of his listeners. Having established our culpability in the matter, he launches into a well-rehearsed tirade, one that we, and most of the crowd, have heard before. The gist of it is that lions have become a big problem, that we are breeding too many in the park, that National Parks and Lion Research do little to control them. And so people are suffering.

Cattle in a boma at Manjelengwa village.

"What are you, Lion Research, going to do about it?"

The bony finger once again gestures in our direction.

"Yes," I reply, "we know this is a problem, but we don't have any control over what wild animals do. By studying their behavior, maybe we can come up with solutions that might help."

The old man looks exceedingly skeptical. He is wondering if we really can be as stupid as we sound.

"Part of the problem," Jane adds, "is that people are not protecting their animals when they are out grazing. They sometimes leave them out in the bush at night. This is when the lions are active. Livestock need to be protected in stockades, in bomas at night."

Everyone nods; they know this to be true. This is how cattle have been protected in this society for millennia.

Our data records show that in this area, the Tsholotsho Communal Land east of the park, has experienced more than 700 predator attacks on livestock during a five-year period. The majority of these took place at night, and most were when animals were left out in the grazing areas overnight.[35] Most families have built strong stockades of teak logs near their homes in which to keep their animals. Many look as if they could withstand a medieval siege. Bomas in this area have much higher walls than in other areas around Hwange, and few gaps for predators to peer through. More important, they are strong enough to keep the cattle in when they panic at the sight of a lion. When the animals break out of poorly constructed enclosures and stampede off into the bush at night, predators get them. Pulling down panic-stricken large herbivores is what lions and hyenas do best. It seems, at first glance, as if sloppy livestock husbandry is the problem here. If people would just use their livestock bomas to protect their animals at night, then lions would not have access to the livestock. Add to that a bit of supervision of grazing herds during the day, and no more lion depredation problem.

However, the solution is not that simple. After twenty years in conservation, I need to stop looking for simple answers. Problems in conservation are often not just biological or environmental; they also have a human, sociological, or economic component. In traditional Ndebele society, children, mostly young boys, would have been responsible for herding the cattle during the day, ensuring they did not stray too far and then bringing them home at night. Today, kids go to school. They are much more motivated by getting an education and a good job in the city than watching the family cows graze. Another problem is that, in the dry season, grazing is scarce; cattle must move further from home to find it in the unfenced rangeland. They also often need to graze at night in order to get enough to eat. If they come back to the boma every night, they lose body condition. Then there is the problem that many of the young people from the community have left this rural community to work in town. There is no salaried work here in the remote Tsholotsho Communal Land. A few people are employed in the tourist industry, but in general, career prospects are nonexistent. The exodus of working-age people leaves a lot of the hard agricultural labor to the older generation. This is particularly severe in the crop-growing season when plowing, planting, tending, and harvesting the food crops the family will depend on in the coming year. In this season, who has spare time to watch the cattle, or energy at the end of the day to go and find them if they don't come home?

It gets even more complicated. Lion predation on cattle is not constant year-round; it peaks in the wet season months. In the rainy season, thousands of ephemeral water holes fill up aross the park and grazing is to be found everywhere. The lions' natural prey disperses widely across the park. As the herds thin out and become less locally abundant, it is hard for lions to find food. Inevitably, hungry lions look across the park boundary and see nice, fat cows. To figure out how cattle are herded in different seasons and how this might affect their vulnerability to predators, we fitted cows with GPS loggers on collars that recorded the animals' position every hour. When we did this, old Mr. Nkomo was quick to point out to everyone who would listen that, in his view, these newfangled inventions actually allowed the similarly equipped lions to home in more easily on the cows. He predicted the GPS-equipped cows were surely doomed. He was only slightly mollified when, after a year, we removed the cattle collars and all the collared cows had survived. When we analyzed the movements of the cattle recorded by the GPS units, the data showed that the distance cattle move from the village correlates closely with levels of depredation by lions and peaks during the wet season months when people grow their crops.[36] We could also see that people herded their cows away from their villages and toward the park boundary in the crop-growing season. This was to keep livestock out of the crop fields, where they would eat and trample the crops. However, the consequence was that, away from the hustle and bustle of the village and closer to the national park boundary, they were a lot more accessible to lions.

The crowd around the cow carcasses is in the mood to discuss the problem further. They don't get much chance to vent their frustrations—national parks rangers and conservation officials hardly ever come here. Tsholotsho is about as far from the seat of government in Harare as you can get, so politicians also rarely turn up, except at election time. As a result, not much in the way of development money makes its way here.

One woman asks, "Why can't there be a fence around the park? We don't mind lions if they stay in the park; we just don't want them here."

We explain that fences are expensive and the government has no money to build one. "In any case," I say, "the wire from the last fence was stolen." Several people shuffle their feet in embarrassment. A veterinary fence was erected along the park's eastern border to prevent the spread of diseases from wildlife to cattle, but it has not been maintained for years. It has become the source of building and snare-making materials for the local people. Only the fence posts and a heavy steel cable remain. Actually, a good electrified wildlife fence would solve a lot of the problems these people face. Predators are not the only species that cause problems. More than lions, people fear a visit from elephants. A couple bull of elephants can devastate a family's crop in a single night. They love nothing better than feeding on the sweet, succulent maize plants in people's fields. Without a good crop, villagers will go hungry. To lose a cow is a hardship, but to lose one's crops is to face starvation.

Over the years our lion research team has had this conversation countless times with people in the communities surrounding the park. There is a real sense of anger here. The people feel they receive little assistance from the wildlife authorities. Unlike in other countries, they are not compensated for losses of crops or livestock to wild animals. Even though they are, in theory, supposed to benefit financially from the wildlife living in communal land, in practice this does not even come close to offsetting the costs they bear. People are all aware of the Communal Areas Management Programme for Indigenous Resources, known by the acronym CAMPFIRE. It was set up in the late 1970s and

early 1980s to allow rural communities to benefit from custodianship of wildlife and other natural resources. The idea was that if people benefitted from sustainably using the natural wildlife resources on traditional community land, it would encourage long-term conservation of the environment and wildlife.

The mechanism for harnessing the value of wildlife could be from leasing concessions to hunting or tourism operators, or harvesting wildlife meat and timber. Unfortunately, despite early successes, the program has largely failed. People see little benefit from the significant monies raised from the leaseholders exploiting their resources. Much of the revenue is wasted on administration and on development projects the people don't feel they have any say in. The program is also

A young male lion shot for killing livestock.

deeply corrupt. In fact, Tsholotsho's CAMPFIRE officer was jailed for embezzling the local program's funds. People know this. Several years ago we surveyed more than 350 households in the village communities to better understand how people viewed wildlife and conservation. One woman pretty much summed up the community's view when she said, "For the *queleas* we blame the Lord, but for elephants we blame CAMPFIRE." (Queleas are tiny finches that form enormous, ravenous flocks that descend, locust-like, on people's crops.)

In the woman's view, the finches could be tolerated as a natural hazard of life. Elephants, however, were the responsibility of the CAMPFIRE program. By implication, because the program was benefiting from the elephants, mostly from payments for trophy hunts on village land, it was also responsible for any damage done by the elephants to people's crops. Elephants and the damage they cause have become a symbol of the failure of the CAMPFIRE program to live up to peoples' expectations. Occasionally, park or CAMPFIRE rangers or trophy hunters will come and shoot crop-raiding elephants. This seldom solves the problem for long. There are many, many more elephants in the park. It does provide a glut of meat for the protein-starved villagers, who quickly butcher the huge carcass with knives, axes, and machetes and parcel out the meat. This is one of the only real benefits people see from wildlife, and nobody here turns down a few kilograms of elephant steak. However, as a benefit, it is infrequent and short-lived and really only reinforces the perception that wild animals, *nyamazaans*, are only valuable as a source of meat, or *nyama*. People get no other compensation for their crop losses to elephants even though they know the CAMPFIRE program sells elephant hunts to wealthy foreigners for a tidy sum. They figure this is not fair. And it isn't.

People's antipathy toward lions and lion conservation follows a similar pattern. They know that lions are valuable to the tourist industry. Game-viewing tourists staying in the luxurious game lodges in the park often pay several thousand dollars a night for the privilege. These are amounts that equate to the annual income of a poor rural farmer. Very little of this revenue makes its way to the rural villagers who are the unwilling neighbors of the lions that tourists love to photograph. Local people also know trophy hunters pay considerable amounts to shoot a lion. This is money that could be, but isn't, used to compensate them for tolerating the lions on village land.

Our impromptu discussion around the stinking carcasses of the two dead cows, recently killed by our study lions, comes to an abrupt close. Old Mr. Nkomo glumly sums up the mood. "In the old days," he says, "our young men would have chased away these lions, or killed them with spears. Nowadays, if we kill lions, we are arrested and go to jail. Who will save us from these lions?"

These were good questions and questions that had no simple answer. The truth is that rural Africans face considerable hardship bearing the costs of living alongside large predators. Though, as old Mr. Nkomo had observed, lethal control of predators by the villagers is illegal. The reality is that, despite legal protection of wildlife, dead cows very quickly equate to dead lions.

Deaths of lions when they leave the safety of the national park and venture into farmland is a persistent problem for the Hwange population, as it is for lion populations across Africa. Hwange is entirely unfenced, and there is nothing to stop movement of wildlife across the park boundary. In total, since 2000, we have recorded the mortalities of more than fifty lions, killed for preying upon domestic livestock around the park. Some of these have been shot by official "problem animal control" teams sent by National Parks, but most have been killed by irate villagers who have lost precious livestock to the hungry cats. The official wildlife-management policy is that lions that kill people's livestock are destroyed by a trained team of National Parks rangers. However, the rangers are poorly resourced, with no vehicles to get to these remote areas and no ammunition for their rifles. They often cannot or do not respond to reports of livestock loss. In these cases, people often take the law into their own hands. As most people do not own firearms, they set homemade snares and traps around fresh lion kills in the hope of catching the culprits as they return to feed. The gin traps, made by local blacksmiths from truck suspension springs, are particularly unpleasant. When set, they consist of two heavy metal jaws that are closed by a strong steel spring when a pressure plate is triggered. They are buried around a fresh carcass, and usually chained to a heavy log. When triggered by the weight of an animal depressing the pressure plate, the heavy-toothed jaws, like those of a prehistoric crocodile, slam shut. The predator's leg or foot is held firmly between the jaws and can often be badly broken by the impact of the closing trap or the subsequent struggle to escape. The trapper shoots or spears the ensnared animal in the morning. Sometimes larger predators fight against the implacable steel jaws and escape with mangled toes and feet. I once treated a young lioness that had been injured in this way.

The manager of Linkwasha Camp, an upmarket photographic safari camp near the eastern border of Hwange, sent me a message that an injured lion was hanging around the camp. Injured lions can be dangerous. Besides being concerned about the lion, she did not want any camp staff or guests attacked.

When I got there I found a young lioness lying under an old ebony tree a hundred meters from the lodge. According to the camp safari guide, she had not moved for several days. Through binoculars, I could see she had an injured forepaw, which she licked and held at an awkward angle. Given that she was close to camp and a potential danger to people, I decided I'd dart her and move her further away. I could then, if necessary, treat her injury. Once I'd darted her and was able to examine the injury more closely, I could see that her foot had been almost severed. It was only attached by a few slivers of skin and sinew, and the bones in the foot were smashed and splintered. An injury like this could only have been caused by a gin trap. She'd obviously been caught while out in the village lands and nearly tore off her foot in her panicked fight to escape the steel jaws. This must have been indescribably painful, and was testimony to her terror at being trapped. There was not much that could be done without significant veterinary intervention, which was not available. I cleaned up the terrible wound and gave her antibiotics and moved her away from the camp. Surprisingly she survived, though she lost her foot and hobbled around on the stump. She never again hunted effectively but managed to get by scavenging from kills made by the rest of the pride. We called her Ginny after the type of trap that had maimed her.

Sixty percent of our records of retaliatory killing of lions are those that are illegally killed in snares, traps, or through the use of poison. It is likely that we vastly underestimate this source of mortality, simply because, unless the animal is fitted with a radio collar, we will often never know it died. Perpetrators of illegal retaliatory killing can be punished with heavy fines or jail time. For this reason, little news of this kind of activity reaches the authorities, and it is likely that a lot of lions just disappear without a trace. Their carcasses are hidden or buried or left for scavengers in the bush. Nobody in the community reports their neighbors to the authorities for killing a marauding lion. Sympathies are very much with fellow livestock owners. The situation is similar in other parts of the world where predators are persecuted for killing livestock. In Scandinavia, for instance, up to two-thirds of instances of illegal killing of wolves go undetected by conventional monitoring. It seems likely that this is also the case around Hwange, so our records are likely to be the tip of the iceberg.

Our records are interesting, though. You'd expect that mortality records would reflect the age and sex structure of the population. We find that almost equal numbers of males and females die in conflicts with people. Surprisingly, though, humans kill many more adult females and young males than mature males and young females.[37] At first this seemed strange. We wondered if it was some artifact of the way we recorded the data. The young male deaths were easy to explain. Young males leave the prides into which they are born, often wandering out of the park to look for a new home. Being inexperienced, they are more likely than adults to kill livestock and therefore to be killed by people. However, adult females are highly territorial stay-at-homes and hardly ever wander far from their pride ranges.

So what was causing them to leave the protection of the park and die at the hands of people? When I looked at the data more closely, an interesting pattern emerged. Spates of livestock killing, and consequently the deaths of adult females, frequently occurred when the territorial male had been killed, often by trophy hunters. It seemed as if disturbance to the lion social structure precipitated the livestock killing. As we'd previously found, the death of a pride male leaves his territory, pride, and cubs undefended against infanticidal rivals. The cubs' mothers, the pride females, do everything they can to avoid new males and thereby save their cubs. They leave the pride home range to hide. Prides on the boundary of the park often leave the park in search of a safe space. But by avoiding infanticidal males, they are exposed to other dangers. Almost all adult females in this situation start to kill livestock, which increases the chances that they will die in retaliatory killings.

This underlines the complexity of human-wildlife conflict. On the one hand, patterns of livestock husbandry and seasonal movements of wild herbivores determine the vulnerability of livestock to lions. On the other hand, the behavior of a social predator determines when lions leave the park and come into conflict with people. The interplay of all these factors, along with social disruption caused by hunting male lions, makes the problem extraordinarily difficult to resolve.

In our village interview survey, we asked villagers how they would like to resolve the issue of human-lion conflict. To our surprise, few demanded that lions be exterminated; villagers tacitly acknowledged that lions had a right to exist. What people overwhelmingly wanted was to be separated from lions by a fence.[38] As it happens, a study that I worked on with Craig Packer and Susan Canney, an Oxford colleague, also showed that lion populations actually fare much better in fenced reserves than unfenced ones. Around half of lion populations living in unfenced reserves are likely to go extinct within the next fifty years.[39] However, fencing large parks such as Hwange is impractical. The expense of doing so would run into several millions of dollars, way beyond Zimbabwe's conservation budget. What is more, fences need to be constantly maintained or they are useless. Abandoned fences across Africa provide steel wire for the continent's poachers. Hwange is home to around 30,000 elephants. Elephants make short work of a fence if they want to get to the other side. Only very strong electrified fences keep them contained. Unless a large international development donor were to come forward to pay for a fence, it is unlikely ever to happen, and donors are not exactly lining up to fund projects in Zimbabwe.

How could we start to solve this problem? It was clear that focusing just on lions would not work; we had to include communities in the solution. First, we had to limit the exposure of domestic livestock to lions. If we could reduce the number of domestic animals killed, we reasoned, we'd also reduce the need to kill lions. Equally, if we could find ways to keep lions in the park, then we'd reduce the chances that lions would come into conflict with people. This sounded a bit like herding cats, and we were not even very sure it would work. As we were considering what could be done, I met with Dr. Guy Balme at a lion conservation meeting. At the time Guy was the director of the lion conservation program at Panthera, an organization founded by philanthropist Tom Kaplan and dedicated to conservation of big cats. He suggested that we might collaborate in implementing something similar to the Lion Guardians program that had just started in Kenya. This program employs Maasai *morans*, or warriors, to protect the lions in the Maasai community areas. This was something of a cultural transformation. Traditionally, morans hunt and kill livestock-raiding lions. They do so using only a spear, the ultimate test of courage. A moran who takes on a lion will bear the scars for the rest of his life, but he will also reap the prestige of doing so. The Kenyan Lion Guardians program employs these would-be lion killers as the custodians of lions, to protect lions from moran hunting parties and to protect livestock from the lions.

We thought that a program loosely modeled on the Lion Guardians could perhaps work around Hwange. It was worth a try, and Panthera agreed to fund it. In 2012 project field assistant Brent Stapelkamp traveled to Kenya to learn firsthand how the Lion Guardian program worked. On his return to Hwange, we put the idea to the traditional chiefs in the two communities bordering the national park. They were enthusiastic. We agreed that each village headman would put forward candidate men and women to be the guardians, not only of their community but of the lions. We soon had six men and two women recruited. What would we call our new team? We christened them the Longshields, inspired by the predominant tribal group around Hwange, the Ndebele. Ndebele warriors would traditionally carry a long rawhide shield, or long shield, for protection in battle. The word "Ndebele" means, "to crouch down or sink out of sight," which is how an Ndebele war party looked to their rivals as they raised their shields before a final devastating assault on their enemies. They became known as "the people of the long shields." Our modern-day Longshields would protect their village communities and livestock from lions and, in so doing, protect the lions from retaliation. Instead of hide shields, we supplied our contemporary warriors with mountain bikes, handheld GPS units, and vuvuzelas. A vuvuzela is the meter-long plastic trumpet blown with gusto by South African football fans to signal their elation or discontent at football matches. The dissonant racket is so earsplitting that international teams at the World Cup football tournament held in South Africa complained the terrible noise distracted their players. The noise of a vuvuzela not only dismays rival football teams, it terrifies trespassing lions. One of the Longshields, David Nchindo, calls his vuvuzela his "machine gun," and indeed, in his hands, it is as effective at deterring lions. David and his fellow Longshields also advise people on strengthening their livestock bomas and assist with the work of doing so. They incessantly nag people to lock up their domestic animals at night.

The project fits GPS collars to lions living in the border areas of the park to make detection of intruding lions easier. When lions are discovered leaving the park, the Longshields leap into action. They warn all the people in the village and urge them to bring their livestock to safety. Once people's livestock is out of harm's way, it is time to persuade the big cats to return to the park. The Longshields gather a group of the more intrepid villagers. Together they blow vuvuzelas, bang planks together, and light firecrackers. They create such a disturbance that the lions, as often as not, turn tail and run back to the

The Longshield lion guardians using vuvuzelas to deter lions from trespassing on village lands.

park for some peace and quiet. This action is not without its dangers; occasionally a lion will take exception to being approached too closely. The chasers can become the chased. Since the Longshield program started, the number of livestock killed annually by lions has dropped by around 50 percent. This is pretty good for a low-tech conservation intervention.

If we don't want to see lion populations across Africa sliding ever closer toward extinction, we have to consider how lions and people interact. The United Nations estimates that Africa's human population will double from one billion to two billion in the next fifty years, and double again by the end of the century. Many more people will need more land, and there will be many more cattle on this land. That is, unless African culture moves away from value systems that promote cattle ownership. More cows and more people mean more scope for conflict with lions and other wildlife. Unless conservationists can find ways to help people tolerate wildlife and for people living alongside wild animals to benefit from their presence, it will become increasingly difficult to conserve large, dangerous animals anywhere except in fenced, protected areas.

# SEVEN

# WHEN LIONS ARE NOT LIONS

A house-sized sandstone boulder sits in the jumbled hills at Bumbusi, on the northern boundary of Hwange National Park. Its top is pockmarked with curious patterns. On closer examination, the patterns emerge as ancient petroglyphs depicting the life-size footprints of animals and people. They show meandering trails of eland, zebras, giraffes, buffalo, lions, and humans. Early people chiseled each image into the red sandstone, using stone or bone tools, at least 2,000 and perhaps up to 4,000 years ago.[40] Similar markings can be found in the rocky overhang to the side of the boulder, a shelter used by the hunter-gatherers who once roamed the landscape that is now Hwange National Park. The artists were clearly superb observers and as familiar with the meaning of each animal track in the dust as we are with the words that form the headlines of the morning newspaper. The painstakingly precise representations suggest that the lives of human and beast were inextricably intertwined—fellow beings whose life cycles fused into one ecosystem. The impression of oneness is never more intriguing than in the places on the rocky canvas where lion and human tracks merge into one another, seemingly part of an intricate dance, perhaps becoming half lion and half person. Significantly, many, but not all, of the carefully engraved lion tracks are five-toed, though

Petroglyphs of lion and human footprints carved into sandstone at Bumbusi.

a lion's paw leaves only a four-toed impression. Given how accurately the tracks of other species are drafted, it is hard to believe this error is anything but deliberate. Perhaps this intermingling of humanoid and feline signified some kind of mythical transformation or was perhaps an acknowledgment of their shared place as adaptable predators at the apex of the food chain.

We seldom see predators in southern African cave paintings and engravings. Where we do, they often appear to have magical connotations, as in the engraved lion at Twyfelfontein in Nambia—a creature whose extraordinarily long tail ends in a six-toed pugmark. At Bumbusi, lion tracks make up a significant proportion of the 700 or so engravings that have been found so far, suggesting that lions were a preoccupation of the people who dwelt there long ago. Early human hunter-gatherers would have shared this landscape with lions. They would certainly have held the big social cats in awe, fearing them and, occasionally, falling prey to them. One only has to spend a night unarmed and unprotected in lion country to recreate that feeling. Perhaps, like the !Kung

Bushmen of the Kalahari, they described all indeterminate nocturnal terrors as "lions."[41] But when opportunity arose, they would also have chased lions off the lions' kills, to harvest a share of the meat just as the !Kung do even today.

Contemporary rural Zimbabweans also have an ambiguous view of lions. A big cat can simply be a lion—known as a *bhubesi* or *shumba*, the four-legged eater of cattle and sometimes people. But sometimes a lion is a servant of the spirit world—a *mhondoro* or *silwane*, an occasionally vengeful messenger from the ancestors, righting the wrongs perpetrated in the community. More rarely, a mhondoro is seen as the embodiment of a powerful chieftain or a long-dead spirit medium. These beliefs are most prevalent in the north, in the ancient landscape perched on the precipitous lip of the Zambezi Valley. The valley—the Zambezi River winding through its belly like a lazy python—is the natural extension of the great tectonic rift that is inexorably cleaving the African continent in two. This hot and sometimes bleak land is dotted with baobab trees—massive, deformed, otherworldly trees seemingly uprooted from a prehistoric land and flung haphazardly across the landscape. A large, bulbous-trunked baobab can be the size of a double-story house. Their gray bark is tortured and gnarled like the skin of an ancient pachyderm. Their branches stand stark against the azure sky in the blinding dry-season heat. The largest ones, sometimes several thousand years old, are frequently sacred to the local people. Legends tell of powerful chiefs whose remains are sometimes interred in the hollow cores of the baobabs and whose spirits inhabit the old trees. To disturb these sacred resting places is to risk meeting with a black mamba, the deadly snake whose venom can kill a grown man in less than an hour. In fact, mambas do inhabit the cool interior of baobabs, often feeding on the bats that also roost there. They do so frequently enough that I have never risked climbing into a baobab.

The mhondoro stalk the supernatural landscape of the Zambezi Valley. The Korekore people who live here believe that spirit lions guard the welfare of the tribe, sending rain and protecting the sacred traditions of the people. They punish those who transgress against the ancestors and those who do not respect the ancestral earth or defile sacred sites. Perhaps not coincidentally, the area seems to be a hot spot of man-eating by lions. In the late 1980s, there was a spate of man-eating incidents here. One of the lions responsible was a young male. This lion invariably killed his victims at dusk, a time when people, meeting on their way home from the fields, enquire of one another, "*Masikati, maswerasei?*" ("Good evening, how was your day?"). With grim irony, locals named this lion Maswerasei. To meet Maswerasei in the gloaming was a poor end to anyone's day. It is not recorded whether the victims had time to make the traditional rote reply of "*Ndasweramaswerawo*"—"My day has been good if yours has been good." In all, Maswerasei killed and ate at least twenty people. He was, certainly in the estimation of the locals, a mhondoro—a lion sent from beyond the grave. They noted that, in every case, his victims had done something to offend the ancestral spirits. They had defiled the land by cutting down sacred trees or broken taboos prohibiting the use of metal containers to collect water from sacred pools. Some had failed to observe the intricate customs designed to ensure social cohesion. One such transgression was the failure to share meat from an elephant hunt with the rest of the community. Because people believed Maswerasei was an avenging spirit lion, they refused to assist the National Parks problem animal control team dispatched to kill him.

We experienced a similar incident just outside Hwange National Park. In April 2014 I received a phone call with the dreadful news that a boy, Elton Ndlovu, had been killed that morning by a lion in Chezhou village, less than a kilometer from the park boundary. The lion responsible was one of our study animals, named Goose. We'd first radio-collared him nearly nine years earlier, in 2005. He'd been one of our core study animals in the Nehimba area, close to the center of the park.

By now he was old, nearly fourteen years of age. Two younger, more aggressive males had displaced him from his territory. Jane had recently caught and fitted him with a new GPS collar. We'd tracked his move-ments over the several months that he was a vagrant, his range drifting ever closer to the park boundary as he avoided other males' territories. A male lion rarely lives to fourteen years in the wild, and almost never dies of old age. The vast majority are killed by people or, less frequently, other lions. If Goose were to live to the end of his natural life and die of old age, this would be a valuable observation for the study. Given that hunters were killing far fewer male lions, the survival of some to old age would indicate the success of the improved management policies. We watched Goose closely and monitored his condition. Surprisingly for an old cat missing most of his canines and seemingly rather decrepit, he appeared to be able to hunt or scavenge well enough to stay alive and in good condition. There was no reason to think he would die any time soon. However, when one day we noticed that the GPS locations his collar was sending had been stationary for several days, we began to suspect he had died. One of the field staff, Lowane Mpofu, and Philani, a park ranger, went to investigate—to collect the collar and, if possible, the skull. That is, if hyenas had not scattered the remains too far. The site was well off the road, and the pair had to walk several kilometers to get there. Confident that they were walking in on a lion already dead, they were so relaxed that Philani did not even cock his rifle. Eventually, through the scrub they could see the gray mound of an old elephant carcass. Their nonchalant approach was abruptly halted when a large, hairy apparition emerged from the other side of the carcass, emitting a terrifying, coughing grunt of warning. Goose was very much alive and well! He had obviously been feeding on the dried-out elephant carcass for several days, which accounted for the stationary GPS locations the collar had transmitted. Philani vanished with an alacrity that would make an Olympic sprinter blink. Lowane, left unarmed and alone to face the lion, wondered briefly what he should do. He remembered that the worst thing one can do is run from a large predator. They can easily outrun a human, and a running form will often trigger the predatory instinct to chase. It was best to back off slowly and carefully. This Lowane proceeded to do. Having extricated himself from the awkward situation, he walked back to the vehicle to find Philani anxiously await-ing his return.

Two days before he killed Elton, Goose wandered out of the national park and, for the first time in his life, confronted humanity. As soon as we noticed that he had crossed the park boundary, we sent warning messages to the local villages through our lion guardian network. Later that day, the lion guardian in the area, Charles Tshuma, organized a crowd of villagers that included Elton's uncle to chase Goose back into the park. In these situations, the "hazing" party does not approach the lion too closely but aims to create enough disturbance and noise to alert the predator that it is in the wrong place and encourage it to return to the sanctuary of the park. This is often a successful strategy, and in this case Goose obliged. Nevertheless, households in the area were warned that a lion was in the area.

The Ndlovu family, having received the warning the previous day, took little notice. Early in the morning of April 4, 2014, Elton's mother discovered she had run out of matches to light the cooking fire. Seven-year-old Elton was told to ask a neighbor for a box of matches. He jog-trotted along the path he had taken many times before, passing through a thicket at the edge of a field. Having obtained the matches, he returned along the same path. But this time, as he entered the thicket, a 200-kilogram cat flung him to the ground.

Andrea Sibanda, one of the project research assistants, saw that Goose's GPS collar was again transmitting from outside the park and went to investigate. Finding his way to the collar's last location, he met with a distraught woman who had witnessed the attack.

"Just a few minutes ago when I was going to the well to fetch some water," she recounted, "I met Elton coming from my neighbor where he had gone to ask for matches to make some fire. I filled the bucket with water and, as I was lifting the bucket to put on my head, I saw

Goose, just after being darted, November 2005.

something like a calf jumping from a thick bush onto Elton. He cried once and some strange noises followed. I screamed and ran for my dear life, and I told my husband about the incident. I don't know what it is, but I suspect it might be a lion."

As the unfamiliar noises of cattle and people filtered through the surrounding bush, Goose had clearly been searching for a place to lie up in the thick vegetation at the edge of the field. His instinct was to hide until nightfall when it would once again be safe to move. He was old and irascible and probably confused and frightened by the new environment he found himself in. He also had little experience with humans, having lived his entire life deep in the national park. We will never know why Goose attacked Elton that morning. As a cantankerous old cat, harried from his range, confused and distressed by the proximity of people, he may have been unusually aggressive. The sudden appearance

Goose, killed after attacking a child, April 2014.

of a boy running through the shady thicket where the big cat was resting may have triggered his instinct to give chase. The attack did not seem to be an act of predation because the big cat did not attempt to feed on the body of the boy. The cold, hard reality may be that Elton was in the wrong place at the wrong time.

The bereaved community of Chezhou village was deeply shocked by the incident, as is natural when a child from a small community is killed in tragic circumstances. Their anguish was in part masked by the stoic fatalism of rural Africa, but also by something less tangible to the casual observer. As was right and proper, the National Parks authority destroyed the offending lion. A crowd of villagers came to gawk at the body of the big cat. But the community did not demand the eradication of all lions, nor did they hold any particular animosity toward Goose, the study lion–turned-killer. In the eyes of some, the attack was not an accident. Many of the village elders muttered that this was punishment meted out to the family for offenses against the ancestors.

The story, widely circulated in the community, filtered back through our local research staff. This is how they later related it to me. The head of the Ndlovu family, Elton's father, was in deep trouble. He had a reputation for dissolute behavior, more often to be found in the local bar than in profitable labor in his fields. He had also failed to provide for his relatives and extended family, a duty taken extremely seriously in African families. As the youngest son, the inheritor of the family's rural home, he had not fairly distributed the family assets, mostly in the form of livestock. This had brought the displeasure of the ancestors and of the conservative community.

In fact, Elton's death was not the first time Ndlovu had had trouble with lions. Curious and increasingly serious interactions with big cats had preceded his son's death. It all started when Ndlovu's stepfather had died several years previously and the burial rites had not been carried out as tradition demanded, and as the old man had wished. This was thought to be at the root of the problem. Even as the mourners had been leaving the old man's graveside, lions had been heard calling from the nearby park. The next morning lion tracks were reportedly found in the fresh earth of the newly filled grave. Over the next few months, lions twice attacked the family's livestock, killing several. In one incident, they broke into one of the family's livestock stockades and killed eight goats, leaving only a single goat kid alive. Becoming fearful, Ndlovu employed a witch doctor to calm the ancestral spirits, but this only seemed to enrage them further. His feline problems reached a crescendo one night after an evening spent sampling the local home brew with one of his cronies at a nearby bar. Transport home after their night of revelry consisted of a two-wheeled "scotch" cart pulled by a pair of donkeys, one of which was accompanied by its foal. It later emerged that the donkeys had been purloined from a neighbor in order to pull the cart loaded with firewood stolen from the nearby protected forest. Indeed, sale of this illegal haul had funded the night's entertainment. Past midnight, on the dusty road to his homestead, much the worse for drink, Ndlovu and his accomplice had fallen asleep in the tray of the cart. They awoke some hours later to the panicked braying of the donkeys. Groggily raising their heads over the low wooden side of the cart, they witnessed a lion pulling the foal to the ground. One of the donkeys, in panic, snapped its flimsy harness and cantered off into the night. The remaining terrified beast dragged the unwieldy cart and its dazed occupants further down the road before a second lion pulled it to the ground. The two men waited out the night, cowering in the back of the open cart, listening to the sounds of lions feasting on the stolen donkey. They perhaps wondered what was in the liquor they had earlier sampled so amply.

In the assessment of the village elders, Ndlovu had not heeded the warnings pointedly provided by the ancestral spirits. This had necessitated a sterner warning. The elders saw the meeting of Elton and Goose the lion as Ndlovu's rightful comeuppance. The loss of a son to an angry spirit lion was testament to the ancestral wrath Ndlovu had brought down upon himself and the people around him. This time there could be no mistake. Ndlovu abandoned his family and home and left the area to live in the nearby mining town of Hwange. Whether to leave behind the memories of heart-rending tragedy or to avoid further retribution, rumor did not say. Yet, even then, lions did not leave him alone. In an incident widely reported in the local papers, two lions entered the residential township of the Hwange Colliery Company,[42] the very place Ndlovu had moved to. Rumor even held that these lions stopped outside Ndlovu's door before vanishing.

Lion attacks on people are exceedingly rare around Hwange. In fact, there has been no other incident of this kind in living memory. Several hunters had been mauled by lions, but in every case it was by an animal they had themselves wounded. Research data we have collected shows lions prefer to stay within the sanctuary of the park. The lions that do leave the park are often on a "hit and run" raid on domestic livestock and usually return quickly to the park after they have fed.

Spates of lion attacks on people are more common elsewhere in Africa, particularly in areas where lions' usual wild prey has become depleted or where encounters between lions and people become unnaturally common and lions learn to see people as prey. One of the most famous cases of man-eating occurred in 1898 when a railway was being built in Kenya between Nairobi and Mombasa through the wilderness of Tsavo. Two male lions terrorized the laborers building the railway, killing, by some estimates, 135 people before being shot and killed by Lieutenant-Colonel John Patterson, the British engineer in charge of the project. A less famous but more serious epidemic of man-eating lions occurred in the Njombe district of Tanganyika (now Tanzania) in the early 1940s. More than 1,500 people were reputedly killed by lions over a ten-year period. The man-eating stopped when the last lion was killed by wildlife warden George Rushby in 1950.[43] In both Tsavo and Njombe, the lion's natural prey had been wiped out by rinderpest, a disease that devastates ungulate populations (animals with hooves). It is speculated that a lack of natural prey caused hungry lions to turn to people as prey.

The "great white hunters" who became famous for eventually shooting the marauding man-eaters may be a thing of the past, but lion attacks on people are not. In an analysis of lion attacks across Tanzania recorded between 1990 and 2005, lion biologist Craig Packer estimates that more than 550 people lost their lives as a result of lion attacks, many of them in southern Tanzania.[44] One of the most infamous of these lions was known to the locals as Osama. Some believe locals named it after Al-Qaeda terrorist Osama bin Laden, though "Osama" means "lion" in Arabic, and coastal East Africa is heavily influenced by its historical links to the Middle East. Osama terrorized the villages along the banks of the Rufiji River between 2002 and 2004. By the time he was eventually shot by the authorities, he had killed and eaten at least thirty-five people. Such was the terrifying effect of his depredations that many people abandoned their fields and farms and moved to the local town. Osama turned out to be a healthy young lion, around three to four years old. He suffered from a tooth abscess, prompting speculation that this might have impaired his ability to hunt wild ungulates and caused him to hunt more easily chewed prey.

Packer's analysis of lion attacks in Tanzania shows that, as in the Tsavo and Njombe cases, once again low levels of natural prey may have caused lions to switch to alternative prey. In an intriguing twist, Packer found that not all wild prey had been eliminated in man-eating areas. Bush pigs were often abundant, possibly because of the taboos surrounding consumption of pig meat by the largely Muslim population. Bush pigs are inveterate crop pests. Many of the human fatalities were people sleeping in their fields in makeshift shelters to protect their crops from them. Packer's team speculates that lions may have been hunting bush pigs, the only available ungulate prey but, on coming upon vulnerable people, attacked and ate them opportunistically instead. This may be how spates of man-eating started, but since lions learned how easy humans are to kill, they have become more specialized.

Similar cases of humans being singled out as lion prey have occurred in Kruger National Park in South Africa. The park runs for 350 kilometers along the Mozambique border. During Mozambique's corrosive civil war, refugees fled the fighting and famine by crossing into South Africa through the national park, thereby avoiding security fences and border patrols along better-manned parts of the international border. In his book *Man-Eaters of Eden*,[45] journalist Robert Frump estimates that as many as 13,380 Mozambican refugees may have been killed and eaten by lions between 1960 and 2005, but this may be overstated. There is no evidence to verify this figure, but it is likely that lions killed many refugees and only a fraction was recorded. Unarmed refugees are extremely vulnerable when walking through a park that is home to around 2,000 lions. It is likely the deaths of many refugees occurred in remote areas and were simply unrecorded. There is no shortage of natural prey for lions in Kruger; it has some of the highest densities of ungulates in southern Africa. However, there is evidence that some Kruger prides learned to hunt and kill people, perhaps even specialized in doing so. When park authorities euthanized one of these prides after a proven attack on a group of refugees, postmortems of the big cats found their stomachs contained human remains, pieces of clothing, and even a wallet containing Mozambican currency.[46] This was not a case of wrong place, wrong time; these lions were actively hunting and eating humans.

Elton Ndlovu's tragic death in a small village on the border of Hwange National Park barely made the local newspaper.[47] The local councillor tried to turn the incident into political capital, roundly blaming lion researchers and National Parks for "taming lions" and "allowing" them to wander out of the park (thus fundamentally misunderstanding the behavior of wild lions and the purpose of research). But even she didn't try too hard, and quickly lost interest in the case as soon as the news reporter headed back to town.

This is in sharp contrast to the story of Cecil the lion, whose death received global news coverage, leading some African commentators to question whether Westerners valued wild animals more than the lives of Africans who had no choice but to live alongside them.[48]

Early humans, such as those who carved lion tracks on the rocks at Bumbusi, might well have been on the menu for large predators. Fear of predation still permeates popular mythology and explanations of the supernatural. Westerners, though, have largely forgotten what it is to be afraid of dangerous predators on a dark night. They no longer live on the edge of the wilderness. Rural Africans in lion country still live with this terror. For them, lions are perilous killers and sometimes even supernatural monsters. It is no wonder they are frequently less enamored of large, dangerous animals than Westerners who live in safe, urban environments.

# EIGHT

# WALKING WITH DYNAMITE

We awoke to the noise of male lions calling up the sunrise, the deep, bass harmony thrumming against our subconscious. Joanne shook me into groggy wakefulness.

"Did you hear that?" she asked. "Those lions are right outside the house!"

The small cottage in which we were staying, a few kilometers inside Hwange National Park, seemed to reverberate with the booming rumble of several big cats roaring in unison.

"Let's go and see who they are," she urged.

By the time we had hurriedly pulled on some clothes and peered cautiously out the door, the cats had moved on, leaving only their fresh tracks. We jumped into the research vehicle and followed their trail, their massive paw prints showing up crisp in the early morning light. The tracks led to Dynamite Pan, a water hole a few hundred meters away, where the four young males were lounging in a dip in the road next a large puddle formed by the previous night's rainstorm. It was not immediately obvious who they were, as they were not from any of the lion prides in this part of the park. They were of an age when young males are forced out of the prides in which they are born and become

144

vagrant dispersers searching for new territory in which to settle. They were relaxed and unconcerned by our presence, and it was clear that they were used to people and vehicles. I began to suspect they were some of the young males who had disappeared from the Ngweshla pride twelve months prior. We later established these were indeed the sons of Pamwe, a majestic lion we'd tagged in 2001 in the east of the park. We'd studied him until the end of 2004, when a trophy hunter shot him. Pamwe's death had led to upheaval and turmoil in the Ngweshla pride. In the absence of a territorial male, the pride had quickly been taken over by a new male, who in turn had evicted all the subadults and mated with the pride females. Pamwe's seven sons and daughter had vanished from the area. As they were close to thirty months old, we suspected that they had not been killed in the pride takeover.

The Dynamite coalition,
November 2006.

The young males rested on the road until the early morning sun
started to filter through the teak forest. Then they ambled down the
track and into the thick wet-season bush to find a shady place to lie up
for the day. We knew they would not move far. That evening, in November 2005, we came back with the darting equipment. I bolted a GPS
collar around the neck of the largest male. We called him Dynamite,
named for the water hole close to where we'd found the group.

Though they were young and not yet in their prime, they were

big-bodied and aggressive. As most males in Hwange are either single-tons or in coalitions of two, few territorial males cared to challenge such a large gang of young males, and they were to dominate the area around Hwange Main Camp, in the north of the park, for the next six years. However, by 2008, just two of the original coalition of four remained, Dynamite and his brother, whom we named TNT in keeping with the group's explosive nomenclature. The other two males in the coalition had been killed in a wire snare set by poachers on the park boundary.

TNT and Dynamite, May 2007.

The remaining duo became known as the Dynamite Boys. They developed a following among tourists and guides and were highly successful territorial males, enjoying longer-than-usual tenure in the Safari Lodge and Ballaballa prides. At least sixteen of their offspring survived to adulthood, which is a remarkably high degree of reproductive success. Two of their sons, Bhubesi and Kakori, are themselves currently territorial males in the national park. The Dynamite Boys were close allies—invincible as long as they stayed together. The battle scars on their faces were a testament to the ferocious fights the pair had with other males who unsuccessfully challenged them for tenure of the two prides. That was until January 2011, when TNT suffered the same fate as his brothers. He was caught and died in a wire snare set by cattle herders illegally grazing their animals in the protected forest bordering the park. His death left the nine-year-old Dynamite alone and vulnerable to challengers for his territory. Dynamite was by now a grizzled old lion. Some would even go so far as to say he was ugly. His nose was black with age, his muzzle almost hairless, and his face scratched and scarred from a hundred vicious skirmishes. He'd become rangy and swaybacked as his prime years slipped away. Over the next year, he slowly lost control of his erstwhile realm, an aging king giving ground to younger pretenders to his throne.

After nearly two decades of studying the behavior of lions in the wild, they still have the capacity to surprise me. Perhaps this is what makes behavioral studies of wild animals so rewarding, and conservation of natural ecosystems so complicated. Just when one thinks one can predict pretty much how lions will behave, they go and do something unexpected. The truth is that wild animals respond to a myriad of behavioral and environmental cues, many of which humans have only the vaguest understanding.

Inexplicably, one day in the middle of December 2011, Dynamite started walking. He left the protection of Hwange. This in itself was not unusual. Lions frequently make forays out of the unfenced protected area, but they almost always quickly return. Dynamite's movement was somehow different. He walked determinedly into the collage of peasant fields and small villages of Africa's rural farmland. Using the data collected by his GPS collar, we could plot the breadcrumb trail of his path on a map and see for ourselves the deliberate arc of his route across the landscape. He walked for thirty-seven days, covering a total of 220 kilometers. He stopped occasionally for a day or two in a patch of thicket between villages, perhaps feeding or resting, before resuming his march. He meandered only a little but kept going more or less northwest, seemingly drawn by the call of some unseen destination that only he could sense. His path led him out of the populated farmland of Hwange Communal Land and over the sparsely inhabited ridges and hills of Kamativi and into the valley of the Zambezi River. On January 15, thirty kilometers downstream from the town of Victoria Falls, he stood on the edge of a steep canyon. This ravine, and the river rushing through its belly, was the most formidable obstacle he had yet encountered on his journey. From fifty meters below came the roar of the Zambezi, where four times the volume of the Colorado River pours through the black-sided canyon of the Batoka Gorge. At this point in its 2,500-kilometer path to the Indian Ocean, the mighty Zambezi, the fourth-largest river in Africa, is forced through a narrow gap carved along a fault line in the iron-hard basalt. The confinement of the great river in the narrow gorge creates foaming rapids separated by eddying, swift-flowing stretches of flat water up to sixty meters deep. These are the white-water rapids that adrenaline junkies from all over the world travel to southern Africa to experience. They are rated between Grades 4 and 6 by white-water kayaking specialists; the British Canoe Union defines a Grade 5 rapid as "extremely difficult, long and violent rapids, steep gradients, big drops and pressure areas." Navigating a Grade 6 rapid is only attempted by the semisuicidal. The names river guides give to these rapids—Gnashing Jaws of Death, Overland Truck Eater, Terminator, and Oblivion—describe their size and raw power, the furious turmoil of the water boiling through narrow portals of hard, black basalt.

To Dynamite, having grown up in the sandy, river-less land at the edge of Kalahari Desert, this must have been a bewildering and frightening sight. Yet whatever force drove him onward caused him to make his way down the sheer-sided ravine to the edge of the swirling water. Like most cats, lions hate to swim. Yet, perhaps after some hesitation, he entered the fast-flowing river below one of the smaller rapids. The current swept him downstream for half a mile before he managed to make it to the safety of the other bank. The positions recorded by his GPS collar suggest he spent several hours resting from his exhausting swim across the river's torrent before continuing with his northwesterly journey, into western Zambia. As a biologist I really have no good explanation for Dynamite's great walk. To my knowledge this is the first time that such an old lion has been recorded determinedly traveling so far from his home range. It may be that with the advent of

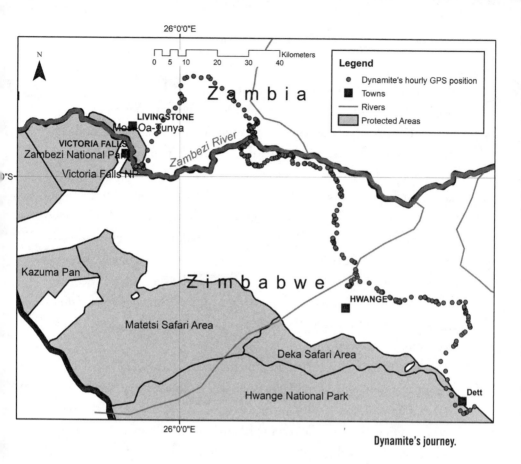

**Dynamite's journey.**

more sophisticated telemetry equipment, this kind of behavior could be recorded more often.

While mature lions very seldom move beyond the neighborhood in which they first settle as young adults, young male lions often travel vast distances from the place of their birth. Evolution has equipped young male lions with the urge to seek out new territories when they are expelled from prides. Dispersal away from the close kin in their natal pride prevents inbreeding and promotes genetic diversity within the population. In populations where there is limited influx of new genes from immigrant animals, population geneticists become concerned about a problem known as inbreeding depression. This is where genetic diversity is lost from the population because close relatives breed because no unrelated individuals have been able to enter the population and pass on their more diverse genes. In large predators, this can eventually result in poor reproduction, which can cause populations to decline. For instance, in the Ngorongoro Crater in Tanzania, the high crater walls created a barrier that effectively prevented the influx of immigrant lions and created a situation in which mating only occurred between close relatives. Scientists studying this population found that male Ngorongoro lions have a high proportion of abnormal spermatozoa, which was thought likely to reduce the chances of conception when the males mated.[49] Loss of genetic diversity can also make small populations much more susceptible to disease epidemics. In populations that are genetically diverse, there is a high chance that some animals will be better adapted at coping with and fighting off infections than others. These animals survive when less-well-adapted individuals die of the disease, allowing the population to eventually recover.

Loss of genetic diversity most often happens when small populations of animals become geographically isolated. For this reason, island populations are often genetically distinct even from individuals of the same species on neighboring islands or mainland areas. Of concern to conservation biologists is the fragmentation of natural habitat by conversion of wild to agricultural land. This often leaves patches of habitat that are effectively islands containing small populations of wild species. These fragmented populations can very quickly lose genetic diversity if dispersing individuals are unable to move between isolated habitat patches.

In the case of lions, the species has lost more than 80 percent of its geographic range over the last 100 years.[50] Between 1871 and 1880, the famous hunter-naturalist Frederick Courteney Selous traveled over much of what is now Zimbabwe and northern Botswana. His expeditions took him through northeastern Botswana as far as the confluence of the Chobe and Zambezi Rivers, through the waterless expanse of what is now Hwange National Park and into the hilly north of Mashonaland in Zimbabwe. He frequently encountered and hunted lions. He considered them to be common across much of the vast area he explored, stating in his book A Hunter's Wanderings in Africa[51] that "All over the interior of

South Africa, wherever game still exists in sufficient quantities to furnish them with food, lions are to be met with, and are equally plentiful on the high open downs of Mashuna country, amongst the rough broken hills, in the dense thorn-thickets to the West of the Gwai River, or the marshy country in the neighborhood of Linyanti." The colonization of these areas by Europeans in the last decade of the nineteenth century brought an influx of settlers, armed with modern heavy-caliber rifles, intent on taming the land for settlement and agriculture. The abundant wildlife that Selous described a quarter of a century earlier dwindled. In a prejudice imported directly from Europe, carnivores were considered to be vermin and killed without mercy whenever they were seen.

Lions in particular quickly disappeared from the areas in which Europeans settled and began to farm and build towns. Wildlife populations became increasingly restricted to patches of wilderness that could not easily be exploited by settlers, parts of the country that were inhospitable, that were waterless, hot, and unsuitable for farming and where diseases of humans and livestock prevented settlement. Even when game reserves started to be established at the turn of the twentieth century, predators were still widely persecuted, as they were thought to suppress "game" populations. One commentator, writing in the *Sunday Times* in 1908, expressed the widely held view that "predatory animals should be confined to zoos of the world; for one cannot expect to confine herbivore and carnivore together without sacrificing the former."[52] This sentiment led to policies that saw predators killed in large numbers in protected areas. In Kruger, the flagship conservation area in South Africa, records show that 3,031 lions, along with countless other predators, were shot, trapped, and poisoned by park rangers between 1902 and 1969.[53] Similarly, Ted Davison, the first warden of Wankie Game Reserve (now Hwange National Park), describes in his memoirs how one of his foremost duties as a game warden was to control predators, an activity he seemingly undertook with characteristic diligence.[54] The result of this wide-scale eradication of predators, along with ongoing conversion of

wilderness to agricultural land, means that much of the lion habitat remaining across Africa consists of small, isolated pockets surrounded by
human settlement and farmland. Many of these small patches of wild
habitat are too small to sustain viable wildlife populations, let alone
large-bodied, wide-ranging species such as lions. As a result, lion populations have disappeared from all but the largest and most pristine areas
of untouched wilderness. In West Africa, recent surveys undertaken by
wildlife biologist Phil Henschel show that tiny pockets of habitat and
the lion populations they support are fast blinking out of existence.[55] It
is only in southern and eastern Africa that significant swathes of habitat remain and where lion populations are more secure. Yet even these
sometimes vast areas are but a small fragment of the range inhabited
by lions in historical times. These remaining lion strongholds are being
constantly eroded.

One important aspect of lion research in Hwange has been to
gain an understanding of the movement behavior of young male
lions as they leave the prides they are born into and seek out a new
territory. Young males are the demographic group in the population
that is most likely to make the journey between isolated populations.
Females are much more sedentary in their movements, with young
females either staying with their mothers or setting up a territory
close by.

To understand the conservation needs of lions in the wider region,
it was critical to understand the behavior of young males as they carried
new genes to distant populations. Few researchers had focused on this
aspect of lion biology. To fill this gap, Nicholas Elliot, a graduate student
working on the project, undertook a study of the behavior of these adolescent cats for his doctoral thesis. It is particularly challenging to study
young male lions because they are often too small to tag with a radio
collar, because their necks grow quickly and the collars soon become
too tight. Just as they become large enough to collar, they leave home,
often without warning. They just vanish. Because their movements are

not monitored, their fates are frequently unknown. Some disappear because they die, others because their movements take them beyond the limited area that biologists can monitor.

Because we knew the prides in our study area extremely well, we were able to predict which young male lions were most likely to be evicted from their natal prides. We monitored them closely and fitted GPS radio collars to them as soon as we suspected they were likely to move. The first movements from the home territory and the protection and security of the pride are fraught with risk. Fatal confrontations with other lions, starvation, or death at the hands of people are frequently the fate of young dispersers. Many make long journeys to find a territory in which to settle. Some make good choices or are lucky; others make bad choices and die. This is the game of chance that all young animals play when they leave the security of home. Some young males made remarkable journeys, which we were able to record using the GPS data transmitted from their collars as they traveled. One such young male was DB—named by a sport-loving member of the field team in honor of English football star David Beckham. DB wandered widely over almost the entire national park. Over a sixteen-month period he made forays into neighboring Botswana and the surrounding farmland. In total he covered 4,224 kilometers—the equivalent of walking in a straight line from New York City to San Francisco or journeying the length of Britain four and a half times.

I learned firsthand from DB about one of the terrifying dangers young male lions face when they leave the safety of home. In fact, I almost got him killed. This happened one day when I was out tracking the lions with Joanne and my friend Godfrey Mtare, an ecologist in National Parks. We found DB early one evening, sitting in the scrub close to a water hole called Kennedy One. It was almost a year since I'd fitted him with his first GPS collar. Its battery had recently gone flat, and we were no longer able to track his movements. As young males move far and unpredictably, it was imperative that we change his collar for a new one. We had a spare collar and the darting gear with us, so we decided we'd catch him and replace the collar.

The size and weight of a young male is sometimes tricky to estimate, so I erred on the side of caution and gave him a large dose of the tranquilizer. The drugs used are exceedingly safe, so there is no risk of an overdose, and it is often better to give a large initial dose than to have to hand-inject an additional dose later if the animal starts to wake up prematurely—a procedure that can be stressful for both animal and researcher! Once we had DB immobilized, it was clear he was very lean, as dispersing animals often are, and therefore much lighter than I had expected. The relatively large dose of the tranquilizer meant he would take longer than usual to come around. This was not a particular concern, as we were not in a hurry and could afford to wait while he recovered. We fitted a new collar and completed the routine protocol of samples and measurements. With the job done, we waited in the late evening dusk for DB to wake. It is my policy never to leave a lion I have drugged until I am certain it has fully recovered. A compromised, half-drugged animal can easily be killed by other lions, hyenas, or even irate elephants or buffalo.

It usually takes around an hour for a drugged lion to come around. However, two hours later, DB had only just begun to stir and showed no indication of getting up any time soon. Under the circumstances, this did not worry me much. What happened next, though, started to get me very concerned. A few hundred meters away, a male lion started to call, soon to be joined by the roars of two others from the opposite direction. These lions—Judah, Jericho, and Job—were the dominant territorial males in this part of the park. All three seemed to be headed in our direction! Territorial males are completely intolerant of young males who are not their own offspring and try to kill them. As a consequence, young males quickly make themselves scarce when mature males are around. However, DB was in no condition to escape the approaching adult males; in fact, he was barely aware of them at all. It was clear we would have to protect the prostrate cat from the incoming males. I parked the research vehicle right next to him and we shone a spotlight around in the hope of providing some deterrence. DB was our responsibility until he was able to fend for himself. The big males soon gate-crashed our small party and almost immediately spotted the trespasser in their domain. Their behavior quickly changed from relaxed to aggressive. It was clear they intended to attack the young interloper who was inexplicably taking a relaxing nap in the middle of their territory. The situation was tense and a little bit frightening for a while. The males would stalk determinedly toward the vehicle; we'd shout at them and bang on the sides of the vehicle. When they got too close, I'd drive toward them to push them off, though one or another would quickly circle around toward the recumbent DB. I even got out of the vehicle a few times and hurled a few clods of dry earth at them. They'd move away and sit watching us from fifty meters away, then circle around and approach from the other side. This continued for several hours. All this time DB slept, oblivious to the mortal danger he was in.

In the early hours of the next morning, the big boys lost interest, finally lifting their siege to continue their nocturnal patrol. DB staggered groggily to his feet shortly afterward and tottered unsteadily to the cover of a nearby thornbush, where he stayed for the rest of the night. We stayed with him. We could not risk him being attacked should the big males decide to return for round two. None the worse for his near-death experience, DB continued to roam widely across the park for another six months. He was eventually killed when he started to eat cattle in the village lands to the east of the park.

In contrast to DB, Dynamite was an old male when he went on his walkabout. With the loss of his coalition partner, he'd been displaced from the security of his territory. Rather than submit to the fate that inevitably awaits male lions too old or too weak to hold a territory, he had chosen to embark on a leonine odyssey into the unknown. He had no previous experience of the land he traversed. We had monitored him from birth, so we knew he had never moved more than a few kilometers outside the protection of the national park. The direction he took was perhaps unfortunate. If he had intended to seek sanctuary in a new habitat and find new mates, he chose badly, for there was no refuge or other lions in the direction he traveled. Had he steered west, he would have entered the vastness of the Kalahari, where perhaps he may have found space to live. Had he steered a few degrees to the north, he may have made it to another island of protection in the form of the Chizarira National Park, perched on an upland plateau 120 kilometers from Hwange.

In fact, a few years later a young GPS-collared male lion took this northern route and achieved what Dynamite had failed to do. This lion was a young male called Kick Junior, the son of Kick, a female from the Caterpillar pride, and fathered by none other than Dynamite himself. His dispersal movement was less surprising than Dynamite's because he was of an age, between three and four years, when all male lions must move away from their mother's pride range and make their own way in the world. Like Dynamite, he ran the gauntlet through the settled land bordering Hwange National Park by rapidly traversing the fields and settled lands at night and hiding in thick bush during the day. Kick Junior was, like his father, also fitted with a GPS collar so we could monitor this movement closely. Day after day, we downloaded the data from the satellite and plotted his new position on the computer screen. Each day he inched closer to the safety of Chizarira, the closest noncontiguous protected area to Hwange. Would he make it or would he fall afoul of snares, poison, or men with guns? After twenty-one suspenseful days, having covered 143 kilometers, he reached the safety of Chizarira National Park. Although almost entirely surrounded by human settlements, Chizarira has its own lion population, and into this population Kick Junior carried genes from Hwange's lions. His journey provided evidence that lions are able to move between isolated patches of natural habitat, even those more than 100 kilometers away. This was an essential piece of the puzzle in the design of networks of protected areas to conserve the species. We've used the movement data we have collected from dispersing male lions, such as Kick Junior, DB, and many others, to help identify the invisible habitat corridors that link up national parks and other formally recognized wildlife areas in the region.[56] If these linkages can be identified and protected by wildlife managers, then there is some hope that lions will continue to be able to move through the landscape and maintain their genetic diversity. This is critical to the long-term survival of the species.

Of course the theory of where corridors and habitat should be and the reality of protecting them are often miles apart. People, and the politicians who represent them, understandably favor economic development in the form of agriculture, mining, and expansion of transport links. These activities are often detrimental to the conservation of wildlife populations and habitat. Africa's economies are on the move and her human population is expanding exponentially; it seems inevitable that development and conversion of wild places will trump conservation.

Nevertheless, there is also significant cause for optimism. First, Hwange's lions are part of a much larger population extending across international boundaries and covering parts of Zimbabwe, Angola, Botswana, and Namibia, known by conservationists as the Okavango-Hwange subpopulation. It is one of only six lion populations in Africa to number more than 2,000 lions. Secondly, this population falls within what is known as the Kavango-Zambezi Transfrontier Conservation Area, often referred to by its acronym KAZA. Protected via an international treaty signed by the heads of state of Angola, Botswana, Namibia, Zambia, and Zimbabwe, KAZA covers half a million square kilometers, an area larger than the state of California. It is one of the largest conservation areas in the world. More than anything else, lions need vast tracts of wild land to survive. Safeguarding the last wilderness strongholds of the species, and preserving the fragile habitat links that join them, is the most critical action conservationists can take to halt declines in lion populations.

Dynamite mating with a Guvalala pride female.

. . .

Having survived the white-water rapids of the Zambezi River, Dynamite struck out across western Zambia, but the trajectory of his walk now veered ever westward. It was as if he was slowly running out of momentum and the gravitational pull of home was drawing him back. The arc of his walk was becoming ever tighter until he was once again on the northern bank of the Zambezi River, five kilometers downstream from the thunderous Victoria Falls. Here the Zambezi plummets into a vertical cataract 100 meters deep, where the walls of the cataracts below the great falls are sheer and impossible to negotiate without ropes and climbing gear. This is also the point at which the river tumbles through three of the largest white-water rapids. Faced with the precipitous gorge and foaming water, his resolve seemed to falter. He stopped walking and his great journey of more than 200 kilometers came to an end. He was also in cattle country, and cattle were the only prey available. The last lions had disappeared from this area half a century previously and the cattle and their owners were consequently unused to large predators. Dynamite feasted on many cows before his cattle-killing proclivities became intolerable. The locals called the Zambia Wildlife Authority to deal with the situation. He was caught and moved temporarily to a captive facility near the Zambian town of Livingstone. When it became clear the collared lion had come from Zimbabwe, a minor international standoff followed. The Zambians wanted to return the lion to Zimbabwe for release into a protected area, a sentiment supported by many who came to hear of the story. However, the Zimbabwean authorities, particularly the government veterinarians, felt that since he had been eating livestock in Zambia, it was possible he had been exposed to bovine tuberculosis, and they did not want this disease introduced into Zimbabwe. For ten months Dynamite's fate was entangled in the copious red tape of several government departments. After much discussion backward and forward, it was eventually decided that he should be released into the south of Kafue National Park in western Zambia. The logic was that lion numbers were low in this area and perhaps he would be able to establish himself here. Maybe this was even the kind of sanctuary Dynamite himself had been seeking when he started his great journey into the unknown. One morning in November 2012, he was once again immobilized by the local wildlife veterinarian and loaded into a crate on a truck to be moved the 120 kilometers north to Kafue. He never made it. Perhaps because he was old, or maybe because insufficient care was taken to keep him cool and hydrated on the long journey north in the November heat, Dynamite died in transit in the crate on the back of the truck somewhere on the road to Kafue.

# NINE

# THE MEAT OF THE ELEPHANT

Tourists visiting national parks in Africa are sometimes a little disappointed in their first lion sighting. Though the animal itself cannot fail to impress, the truth is that, for much of their day, lions do very little. In his intensive study of lion behavior, George Schaller calculated that over a twenty-four-hour period, lions are typically inactive between twenty and twenty-one hours; they walk (usually slowly) on average for two hours and spend the remaining one to two hours hunting, feeding, socializing, mating, or fighting.[57] During daylight hours, lions loll around as languid as toffees on a hot pavement, and after the first ten minutes of watching them, they are just about as interesting. This air of somnolence conceals their importance as apex predators in the savannah ecosystems in which they live. There is nothing quite like the exhilaration of watching lions metamorphose from lazy layabouts to muscular, active predators.[58]

Early on in the study of the Hwange lions it became clear that, without an understanding of how the local ecology was influencing lion behavioral ecology, it would be impossible to make any wide-ranging recommendations about how the lion population could be conserved and managed. Added to this, every ecosystem is different, driven by varying climatic conditions, soils, and assemblages of prey, as well as by differing levels of conservation management. Much of the work on lion biology and ecology had been undertaken in environments where lions were relatively easy to study, such as the Serengeti plains. Hwange is part of the Kalahari system, which is arid, with erratic rainfall and has an ecosystem of savannah woodland and dense, scrubby, dry forest. This is an environment where lions are sometimes tricky to find, let alone study. Research on predatory behavior is more complicated than it might at first appear. Because lions spend long periods conserving energy between bouts of hunting and feeding, if you want to study patterns of predation, you either have to be very patient and have a lot of time on your hands, or you must find alternative ways to collect the data you need. It is also challenging in the heavily wooded habitat of Hwange, where only about 5 percent of the park is open savannah grassland. It is almost impossible to follow lions into the wooded areas in a vehicle, and doing so on foot would be impractical, not to mention dangerous. If one only recorded lion hunting behavior in the open, accessible areas, this would introduce significant bias into the study. Our field team, initially led by Zeke Davidson and later by doctoral student Moreangels Mbizah, used GPS collars to track the lions. These instruments record the lion's location at predetermined time intervals (every hour in our study) and transmit the data to the researcher via a radio or satellite link. When lions make a kill, they often spend considerable time at the same location, feeding and resting. This can be a few hours for medium-size prey such as a kudu or zebra, and sometimes as much as a week when feeding on a giraffe or elephant carcass. By recording when the lions remained in the same location for several hours and investigating these locations on the ground, we were able to get a good idea of what the lions were killing or scavenging. In this way, we were able to paint a picture of what prey species the Hwange lions relied on, how this changed with the season, and how lions used different parts of their home range. Ultimately we were interested in what factors determine how lions use the landscape and what might constrain the number of lions that could subsist in the national park.

We undertook a more detailed study to understand exactly how often lions make kills, how often they feed, and, importantly, how often they hunt but fail to make a kill—information we could not glean from the GPS collars. To do this, we fitted lions with, along with GPS collars, small custom-built activity loggers. These are rather similar to exercise trackers marketed by the leisure industry for people trying to get fit or lose weight. They detect motion in three axes and with suitable calibration can be used to determine the lions' activity on a second-by-second basis. We found that lions attempt to hunt almost every day, on average making five chases a day. However, they are only successful 12 percent of the time, or one out of seven of the hunts they attempt.[59] This is a rather similar success rate to lions observed hunting in Etosha National Park in Namibia but lower than the success rates of lions in the much more open, prey-rich habitat of the Serengeti.[60] Lions have a feast-or-famine lifestyle. When they snack, they snack big and often, feeding every second day, but in times of famine, they can go days without eating. In exceptional circumstances, they can go without food for two weeks.

Lions are particularly specialized and indeed seem to prefer to hunt the large- to medium-size herbivores that once formed abundant herds across savannah Africa. It was once thought that lions evolved to form social groups in order to subdue and kill the large prey animals they hunt. That is, until it was observed that a single lioness is quite capable of pulling down a zebra or wildebeest on her own. A pair of lionesses can take down a buffalo or giraffe, and their rate of success in doing so is no lower than the hunting success of larger groups. There is no need for a duo of huntresses to share their kill with other hungry lionesses. The more mouths there are to feed, the less food each individual gets. There is no hint of sharing or cooperation when lions feed. On a kill, it is every lion for itself. They growl and swat at each other in the fight for a prime spot at the table. There are perhaps more compelling explanations for the evolution of group living among lions. There might be an advantage in having the rest of the family to dinner when it comes to defending the carcass against other lions or ravenous spotted hyenas. On several occasions I have seen small prides lose their hard-won kills to cackling gangs of hyenas. Larger groups of lions are better able to see off would-be thieves. It is also likely that communal protection of cubs in a crèche and cooperative defense of territory against neighboring lions might explain why related lions band together.

That is not to say lions do not hunt cooperatively. One evening I was watching the Safari Lodge pride; six large females were lying in the scrub on the edge of the open valley, listlessly eyeing a pair of zebras several hundred meters away. Suddenly they all flattened themselves to the ground, each cat dissolving effortlessly into a clump of grass. They vanished so completely that, scanning with my binoculars, I could glimpse only the occasional black-tipped ear. A small herd of wildebeests was ambling toward the pride, oblivious of the presence of the cats. The stalking lions had instantly fanned out in a line to blockade the path of the advancing herd, which would surely spot them in the next few minutes. Then Stumpy Tail, unnoticed by the wildebeests, flanked stealthily around the herd until she had them perfectly set up for an ambush. Surely she would now pounce. In a moment of hesitation, perhaps she had missed her opportunity. She melted into the sun-bleached grass while the herd passed her by, marching on toward the ambush line of invisible cats. I held my breath, waiting for the lions to rush in and snag their dinner. The unseen lions waited too. The wildebeests stopped. Unconcernedly they started to graze, creating the illusion of a peaceful evening tableau. That illusion almost immediately shattered when, like a well-drilled sports team responding in unison to a predetermined signal, the five lionesses stood up. As the lionesses materialized from the grass, panicking wildebeests whirled around in retreat, hitting a gallop in a few seconds. Their chaotic flight from the five brought them right past the small grass clump behind which Stumpy Tail lurked. She was a blur as she hit the nearest wildebeest cow, bringing it down in a tangle of legs and dust. Her claws embedded bloodily in its thick hide, her jaws clamped on its throat. A few muffled bellows and futile thrashing of legs followed. It was all over in less than a minute. By the time the rest of the pride sauntered up, the unfortunate wildebeest had stopped moving. Stumpy Tail had already started to open up its belly to access the nutrient-rich vital organs. The cats crouched down and started to feed alongside her, their faces soon bright with warm, congealing blood, as red as the deepening sunset behind them. In retrospect, the hunt had been almost perfectly choreographed. Stumpy Tail, the most experienced hunter, had taken point position, choosing the killing ground to perfection. The five other females had made little attempt to pursue the fleeing wildebeest herd. Their function had been as beaters to create panic and drive the terrified herd into the careful ambush laid by Stumpy Tail. This was not a chance success. I saw the pride employ exactly this tactic on other occasions.

Lions use different hunting techniques for hunting different prey species. A pair of lionesses called the Backpans pride puzzled us when we looked at the data we had downloaded from their collars. We could find no evidence of kill sites, yet they were healthy lions that clearly fed regularly. To feed, they must be making kills. Jane eventually figured out what they were doing. The pair of lionesses was hunting during the day; their specialty was catching warthogs. Warthogs are alert and speedy—much faster than a lion, so running them down hardly ever works as a hunting strategy. The Backpans girls had a much better method. Warthogs use underground burrows for protection at night when, in the dark, their excellent eyesight and rapid escape speeds are less of an advantage. The lionesses had learned to lie in wait at the entrance to the hog burrows in the early morning, just as the warthogs were becoming active. If the cats were lucky, they could snag an emerging pig that had breakfast on its mind and was not paying attention. If that failed, they reverted to less subtle means. In a flurry of sand, like large dogs excavating a bone, they would dig the pigs out of their shallow burrows. This means of hunting is not without risk because, when heading for bed, a warthog reverses into its burrow, defending the opening with razor-sharp tusks. Indeed, one of the Backpans females had a massive scar across her flank, likely from just such an encounter. We seldom found evidence of their kills because, having caught a tasty hog, the two lionesses would quickly and completely consume the carcass, leaving nothing but part of the skull and tusks and sometimes the hooves. They spent little time hunting or eating at any one location, so it was hard for us to pinpoint their kill sites from the location data we collected. This changed when the Backpans pride grew larger and the females had more mouths to feed. They then switched to hunting larger prey.

Other prides learn and refine particular hunting techniques for catching the prey that might either be abundant in their ranges or of a suitable size to feed the pride. The Shakwanki lions, a large pride of ten females, excelled at killing giraffes. They would ambush the ungainly creatures when they were in thick sand. Once on the move, a giraffe's long legs and huge stride allow it to cover the ground rapidly, with the gait of an unwieldy rocking horse. At top speed it can easily outrun a sprinting lion, and lions do not have the stamina for a long chase that would be necessary to tire out a giraffe. But reaching top speed takes a few strides for such a large animal, especially when struggling to get traction in the thick Kalahari sand. This slight delay allows the Shakwanki girls to leap onto the tall animal. With several lions clinging to its back and neck, the giraffe is quickly toppled. Once down, with several lions on top of it, it cannot get back up.

Lions use a similar mobbing strategy when they hunt elephants and buffalo. The game plan is for one or two lions, while avoiding the swinging trunk or scything horns, to jump on their quarry's back. Holding on with their razor-sharp claws, they ride the bucking animal like rodeo cowboys. This slows the victim down enough for the rest of the pride to gain purchase. As any rugby player will tell you, even the strongest opponent cannot stay on his feet under a scrum of determined bodies.

The Ngweshla pride hunting
a juvenile elephant.

Once off their feet, belly and throat exposed, prey animals have little chance of getting up again. Male lions quite often clinch the outcome of these battles by helping to pull down the larger prey animals. The addition of a 200-kilogram male to the fray is usually enough to knock a buffalo bull or young elephant to the ground. I have seen them saunter up to a buffalo that the lionesses have caught but are struggling to pull down and with one wrench simply pull it over. It is almost as if they are saying, with masculine arrogance, "Don't worry, I'll take it from here, ladies." It is something of a fallacy that male lions do not hunt, though if they are with a pride, they are often content to let the lionesses do the heavy lifting and, when the job is done, claim the best seat at the dinner table. Nevertheless, they are efficient hunters in their own right. In fact, male lions often seem to specialize in catching larger prey. I once saw two males deftly pull down an adult bull giraffe weighing in at more than a ton.

Of course hunting large, dangerous prey animals is not without its risks. I've seen a young male lion tossed several meters into the air by a

buffalo. He survived the encounter but was obviously bruised and sorry
for himself afterward. Other lions are not so lucky. We once found the
remains of a lioness trapped under a giraffe carcass. It seemed that when
she and her pride had pulled the giraffe over, she had been caught un-
derneath as it fell and had been crushed to death. A significant number
of the several hundred lion skulls I have examined over the last twenty
years show signs of traumatic injury, much of it likely to have been
caused while hunting. Many lions have broken teeth, and a significant
number show signs of factures on the jaw and cranium that could only
have been caused by a kick or swipe from a large prey animal. These
bone injuries have frequently knitted and healed over. Given that lions
must subdue frantically struggling and often aggressive prey animals several times their own size, using only claws and teeth, it seems inevitable
that such injuries will occur.

In Hwange October is often the best time to see lions hunting,
just before the rains come to revive the dusty bush. The scorching-hot
end of the dry season is the death time for many herbivores. Daytime

temperatures nudge 40 degrees Celsius (104°F). Water is scarce, few leaves remain on the starkly bare trees for browsers to eat, and the remaining grass crumbles underfoot into dust. Except for the carnivores, most species lose condition at this time of year. Many, especially the old and young, are weak and vulnerable to predators. Others simply collapse and die of malnutrition or dehydration. Scavengers, including lions, quickly consume their bodies. It is nature's way of weeding out the old and infirm. For carnivores, it is a time of plenty, a time to feast to bursting in preparation for leaner times. The lions are sleek and well-fed and the lion cubs waddle from meal to meal like footballs with legs.

Lions have learned that in the dry season they just need to sit and wait near the water holes to pick off the stragglers as thirsty herds come to drink. Analysis of lion movement and kill data by my colleague Marion Valeix shows that lions make a high proportion of their kills within two kilometers of a water hole. For lions, the water holes are the prey-rich larders in their territories—so important that lion range size is determined by the density of water holes. Water holes are close together in small ranges, widely spaced in large ranges.[61]

When Ted Davison, the first warden of the park, arrived to take up his post in 1928, there was no surface water in the park for much of the dry season. As a consequence, there were very few animals. He made it a priority to provide a reliable supply of water within the park for wildlife. This was achieved through drilling boreholes next to existing seasonal water holes or pans. The water was pumped into cement troughs, which overflowed into the natural pans. This technique allowed many of the water-dependent species to live permanently within the protection of the park. Today the thump of a diesel engine pumping water from deep underground to supply the water holes could almost be the heartbeat of Hwange in the dry season.

The rains in early 2011 had been particularly poor, and by the end of the year, the park was bone-dry. The guides from Makalolo Camp had been reporting that Ngweshla pride had been killing elephants for several weeks. The inevitable dry season slaughter had started. This is the perfect time of year for catching and collaring the animals that we need to monitor. At this time of year, one can find lions lolling in the shade near almost all the large water holes, digesting their latest meal.

We'd driven a few hundred meters off-road in the Land Cruiser through the open forest to find the Ngweshla pride. In the narrow shade thrown by the broad trunk of a large umtshibi tree, the lions were sprawled in a long line, making use of every inch of cool sand. They panted in the heat, their chests heaving rhythmically. As we watched them, we sweated in the hot, still air. Our discomfort was intensified by a swarm of mopane flies attracted to our salty sweat. They are actually miniature, stingless bees, which sometimes seem to have evolved especially to torment field biologists. In the parched, dry bush, they harvest all available moisture, from one's eyes and inside one's nose and mouth. They burrow into your hair and explore your ears. The relentless flies also covered the lion's faces. Occasionally one of the cats would sneeze violently as one of the little pests crawled up its nose. As the day cooled off, the lion cubs tired of the siesta and started to stalk the twitching black tips of the adults' tails. Like well-fed, energetic youngsters anywhere, they wanted to play. The grown lions hardly moved. Our hopes that the pride would hunt were not likely to be fulfilled. It was clear they were going nowhere after gorging themselves the previous night. We headed back to camp to return the next morning.

When we returned just after sunrise, they had not moved far. But this time they were anything but sleepy. A terrified juvenile elephant, about four years old, came tearing through the low scrub, closely pursued by a dozen tawny forms. The whole pride was there, and even the year-old cubs seemed to be getting in on the action. The mob of lions pursued the elephant for several hundred meters. Occasionally, one of the lionesses or young males would jump on its back, only to be thrown off. Eventually Jericho, the big male lion, brought the chase to an end by pulling the elephant onto its haunches. The elephant almost

Rescuing a baby elephant.

disappeared under a blanket of lions. What followed was not pretty. Lions are good at killing ungulates, suffocating them swiftly by clamping the windpipe closed. An elephant breathes through its trunk, so this does not work. Unless the lions get a grip on the trunk and close off the airways, as some lions learn to do, the elephant dies from blood loss as the lions start to feed. In the case of a large elephant, this can be agonizingly slow. Nature is often brutal.

People are sometimes surprised to hear that lions kill elephants. Actually they do so only rarely in most places. Adult elephants are too big even for a determined pride of lions, and the tightly cohesive elephant herds aggressively defend the more vulnerable young elephants. Elephants hate lions, and they will often chase them, charging after them, trumpeting shrilly. This usually seems more of a warning demonstration, as the lions mostly don't seem too worried and only scatter and retreat a short distance. The lumbering giants have little chance of catching the agile cats. In Hwange, elephants have every reason to react aggressively toward lions. Here lions regularly kill and eat elephants, with the pachyderms making up between 10 and 20 percent of the lions' diet in any one year. Our data shows that elephant killing happens most often in the late dry season of very dry years. When water is scarce and elephant herds move long distances between water and feeding grounds, young elephants become exhausted and lag behind the herd—easy pickings for a pride of lions.[62]

On several occasions we have found young elephants trapped in the cement water troughs that receive the water pumped from boreholes. They fall in when the herd is drinking and are too small to get out. Rather than leave them to die in the troughs, we always try to rescue them. They are usually babies, so they are small enough for two or three people to wrangle. Once out, they at least have a chance of reuniting with their family. A couple times I have seen their mothers come back for them. In one case, an angry cow elephant charged us as we were lifting a calf out of a trough. She was clearly the mom and objected most strongly to us manhandling her calf. We dumped the calf out of the trough and ran for the vehicle, speeding off as the cow and calf reunited. This was one of the happy endings. More often, the calves just seem lost and bewildered. Their instinct is to follow anything large. One time a young calf we'd rescued became completely fixated on the research jeep I was driving. When I tried to drive away, it followed the vehicle, just as it would its mother. Nothing could persuade it to leave its newly adopted protector. It was heartbreaking to have to speed away and leave it behind. Unfortunately, the park rules did not permit us to intervene any further.

On another occasion we got to see firsthand the fate of these young, orphaned elephants. This happened at Nyamandlovu Pan, a large water hole whose name translates, somewhat ominously, as "meat of the elephant." Joanne, Jane, and I found a calf stuck in a deep ditch that had been dug around the diesel borehole pump—ironically, to stop elephants from damaging the pump's piping. The little elephant was jammed into the ditch. We could see from the elephant tracks around the pump that the herd had milled around for some time, trying to rescue the stricken calf. It was less than a year old, but it took all three of us and the National Parks pump attendant to rope and heave the elephant out. The poor animal was desperately thirsty, so we fed it from our water bottles. There was not much else we could do; we could only hope it would find its mother. Just as we were getting ready to leave, the young elephant headed off into the bush in the direction of a small herd that was approaching the water hole. We hoped they were his family group. The next afternoon we were tracking the local pride of lions. We found them well-fed in the shade of a scrubby blue bush, close to Nyamandlovu Pan. Close by were the remains of an elephant calf the lions had killed during the night. We could not tell if it was the same elephant we had rescued, but in our hearts we knew it was.

The word "carnivore" is derived from the Latin word for meat, "carne," and translates literally to "devourer of flesh." The big cats are entirely carnivorous, adapted to a diet exclusively composed of animal protein. An adult male lion can eat as much as thirty kilograms of meat at a sitting, roughly 15 percent of his body weight. Young lions eat proportionately more, with yearlings consuming as much as 25 percent of their body weight in a single meal.[63] After gorging themselves at a kill, the stomachs of these young cats protrude uncomfortably like overinflated balloons. To survive, an adult lion needs on average between five and eight kilograms of meat a day, or nearly 3,000 kilograms of meat a year. This is the equivalent of eating around thirteen wildebeests or five buffalo a year per lion—or, for the urban-adapted reader, 66,000 Big Mac burger patties. A large pride of lions needs to catch the equivalent of 150 wildebeests per year, equating to 30,000 kilograms of meat. This vast protein requirement is bad news if you are a wildebeest, zebra, giraffe, or buffalo, the species most favored by lions. It is also bad news if you are a livestock farmer and hungry lions from the nearby national park pay you a visit.

The need to consume significant quantities of meat is, fundamentally, what makes conserving large predators so challenging. A population of lions can only exist if there are sufficiently large populations of wild herbivore species on which the lions are utterly dependent for food. These large, herd-living herbivores, in turn, need huge tracts of wilderness to exist at all.[64] In modern Africa, more and more land is being converted to farmland and space for wildlife is becoming increasingly limited. No space equates to no herds, and this in turn equates to no predators. This is the real threat that lions face.

Collaring one of the
Mpofu coalition.

# TEN

# GAME OF THRONES

Mpofu, the ragged old cat we were watching, was twelve years of age. Lions rarely live this long in the wild; their lives are hard and the fight for existence is brutal. The thick, black mane he was famous for was as luxuriant as ever, but he was skeletally thin and his eyes had lost their yellowish fire. Every rib was visible and the vertebrae of his spine protruded through his skin like a line of tombstones. His condition had deteriorated noticeably since I had last seen him, and now he could hardly walk. A lion that can't walk can't hunt or even scavenge, and he was slowly dying of starvation and the raging infection in his leg. Four months before, a challenger in the Mangisihole coalition had bitten him in the knee joint of his hind leg; the wound had never healed properly. It is the national park's policy that in such cases of natural injury, nature is allowed to take its course. Mpofu's injury in a fight with rival lions was, for a lion, as natural as it gets. There could be no intervention to save him. Still, I could not help feeling a sense of sadness for the suffering of a lion I'd known and studied for almost a decade. He had been the dominant male in this area for just over nine years. He'd fathered more than fifty offspring in six different prides; the genes for this magnificent black-maned male lived on in several generations in the study population. As we watched the old lion lying quietly, his massive head resting on his paws, seemingly resigned to his end, I thought back to the first time, ten years before, when I had seen him as one of a coalition of four arrogant young males.

In October 2000 I was searching for new lions deep within the national park at Shapi Pan, the very place where twenty-eight years before, Warden Len Harvey had been killed by a lioness. The old National Parks camp had been abandoned shortly after the terrible incident, but some remains of the staff houses could still be found in amongst the teak trees. Perhaps because of its gruesome history, Shapi always gave me an uneasy feeling. Lions here seemed to be more self-assured and bolder in their interactions with humans, possibly because they saw few people so far from the main tourist areas. A few years before, lions had tried to get into the corrugated iron shack provided as accommodation for the National Parks laborer who looked after the diesel pump supplying water to the Shapi water hole. The terrified pump attendant had spent the night cowering in his flimsy hut as the lions attempted unsuccessfully to gain access via the tiny window and poorly fitting door. Since then, no one had lived permanently in this part of the park.

In the muted red light of the dying October day, a quartet of three-year-old lions had swaggered out of the scrubby bush and onto the calcrete pan to greet the oncoming night. The setting sun turned them a fiery gold, their manes seeming to flicker in the lengthening shadows. I managed to snap off a few blurred photographs before they crossed the open area. I tailed them down an elephant path into the gathering gloom. I could not follow them through the thick scrub into which they disappeared. I did not see them again for another four months.

When I did finally see them again, they were feeding on a buffalo at Mpofu Pan, and I was able to catch and collar two of them. The newly tagged lions were named the Mpofu coalition, after the water hole. "Mpofu" means "eland," a large African antelope, in the local Ndebele language and is also a common Ndebele surname. We captured the two

Two lions from the Mpofu coalition lie darted, while the other two continue feeding, unconcerned.

other males in the coalition later in the year and were able to closely monitor the coalition's movements over the next few years.

By following the fates of known individuals in the Hwange lion population for close to twenty years, we have traced their intricate relationships. In his novel *Game of Thrones*, George RR Martin plots the murderous intrigues of several fictional dynastic families, set in a fantastical and brutally medieval world. Martin could easily have modeled his storyline on Hwange's lions, for their social lives are as convoluted and the interactions between rival families no less violent. Our research has recorded remarkable stories of alliance, collusion, fratricide, tragedy, and conquest that rival the battles and intrigues of the Starks and Lannisters. The difference: lions do all this not for greed, lust, or political power but because of an evolutionary drive to pass on their genes to the next generation. To do this, they must acquire and defend prime real estate that will provide the food resources to sustain them and their mates. Their intricately cooperative social system has evolved to allow individuals to access territory, food, and mates, and to successfully raise offspring. These complex social networks are precariously balanced and easily knocked off course by the disturbance of human actions. The death of a key individual in a snare or at the hands of a hunter or poacher has cascading effects on the remaining lions. The death of one lion frequently results in the death of several others through starvation, infanticide, or conflict with other lions or people. Hwange's own Game of Thrones was played out by two strong coalitions of males that the project monitored almost from its inception. The fate of the Mpofu coalition and its descendants, and their interactions with the Mangisihole coalition, underlines just how complex and convoluted is lion social behavior.

Over the next few years, the Mpofu coalition grew from the lanky subadults I'd first seen at Shapi into magnificent full-bodied males with the full, shaggy manes of true Kalahari lions. Three of them had dark manes that darkened further as they aged. The fourth, the one that had awakened prematurely after the first time I collared him, was noticeably blonder; naturally we had to call him Blondie. The largest dark-maned male was called Mpofu, and the other two we knew as Thirty-Three and Ninety-Five, the radio frequencies of the collars we had fitted them with. The core of their home range covered the prey-rich plains around Hwange Main Camp, where the tourist traffic is clustered. Consequently,

Mpofu.

the Mpofus became completely relaxed around park visitors' vehicles. Blondie in particular was so laid-back that he would sprawl, sleepy and immoveable, on the game-viewing tracks and force tourist vehicles to drive off the road to get around him. When I recaptured him to replace his standard radio collar with one that would collect GPS locations, I literally drove up to him and darted him as he lay on the road. Even then, he hardly bothered to move. It was the easiest lion capture I've ever done. Although the center of their range was Main Camp, their movements also covered an enormous swath, with a range size of 1,500 square kilometers, an area equivalent to Greater London. They appeared to wander at will, consorting and mating with at least four separate prides of females, seemingly because there were few other large males to constrain their movements. Inevitably, though, their peregrinations took them beyond the sanctuary of the national park.

In June 2003 the Mpofu coalition wandered into a hunting area to the northeast. Jane was out searching for them when she realized that

the radio signal from the collar of Blondie—the naive, trusting one—was coming from a hunting camp. When she got there, she followed the signal to a shed where slain animals were skinned. Jane marched into the shed and claimed Blondie's collar from the astonished camp staff. I

Blondie, one of the
Mpofu coalition.

felt deflated when she reported that Blondie was dead. He had become a favorite lion. It was hard to see how it would be much of a challenge to hunt him, given that he was completely habituated to humans. It must have been a bit like shooting a dairy cow grazing in a field. His brother,

Thirty-Three, was shot a few days later. Bullets claimed a third brother, Ninety-Five, toward the end of 2003. The horrible year saw eleven lions killed, nine of which were collared study animals.

With his three brothers dead, Mpofu was the last of the four-strong coalition. He now rarely left the national park. Perhaps he had been present when his brothers had been shot and he'd learned the dangers of feeding on baits in the hunting areas.

Over the decade Mpofu was monitored as a study animal, he was captured and collared seven times. Although he was not particularly bothered by vehicles and could usually be approached closely, he seemed to sense when we were after him to change his collar. He never ran away but would quickly find a spot, usually deep in a dense thorn-bush, where he seemed to know we'd have difficulty accessing him. He was particularly adept at keeping enough vegetation between himself and the dart rifle to prevent a clear shot. We had to get equally cunning. On one occasion, Jane distracted him from a vehicle on one side of his chosen bush while I waited in a vehicle on the opposite side of the bush with the dart rifle at the ready. With his attention focused on Jane, he carelessly revealed his shoulder through a tiny gap between the thorny branches. I was able to snap off a shot, and twenty minutes later we were bolting on his fourth collar.

As a singleton male, Mpofu would normally be vulnerable to a stronger coalition of lions displacing him. Luck, or rather circumstance, was on his side, in that the intense trophy hunting in the years up until 2003 had left the northern part of Hwange depleted of both mature and large subadult males. Because there were so few male rivals in the population, his tenure as a pride male was not challenged. By the end of 2006, Mpofu, now nine years old, had sired nineteen cubs in the Balla-balla and Guvalala prides.

But that year, Mpofu was dethroned. It was a case of "live by the sword, die by the sword." The young Dynamite coalition was as merci-less as he and his brothers had been when they had taken over the area six years earlier. The Dynamites soon mated with the young females in Mpofu's old prides and banished Mpofu's four three-year-old sons. All vanished; at the time, we did not know whether the Dynamites killed them or whether they had dispersed in search of a territory of their own. Mpofu himself, outnumbered and outgunned, headed to the Masumamalisa Valley in the east. Here he took over a small pride whose range encompassed a water hole and popular tourist picnic site known as Kennedy 1. We knew them as the Kennedy pride. They were a skittish group of females, led by a duo of ancient lionesses. The pride had a hard life right on the boundary of the park, adjacent to an active hunting concession. Their pride males were frequently shot by hunters, and as a consequence, they seldom managed to raise any cubs. At the time Mpofu was ousted from his former prides, the Kennedys had no pride male and the opportunist soon established himself as the new territorial male.

In late 2007, in Mpofu's tenth year, we heard that three young males had moved into the Kennedy area and were frequently seen close to the Kennedy pride. We guessed this would be the end of Mpofu. Once again, numbers were not on Mpofu's side; the young trio would quickly kill or displace the old boy. Oddly, though, during the months we observed them, the three young males and Mpofu appeared to be tolerating each other. This was peculiar behavior, as a territorial male would not ordinarily behave amicably toward potential rivals. One evening we found the young males feeding on an elephant carcass close to the Kennedy 1 water hole. I was able to dart one and fit a radio collar to him. While we waited for the young male to recover from the anesthetic, Jane consulted the ID cards the project keeps for each lion in the study. It turned out that these young males were in fact Mpofu's sons from the Guvalala pride. They had survived the takeover of their mothers' pride by the Dynamite coalition. This explained the lack of aggression Mpofu showed toward them. Gradually, the one old and three young males spent more and more time together and, just as if they were coalition partners, defended and shared the pride range and mating opportunities

Mpofu (left) and his
son Jericho (right).

in the pride. Of course this made perfect sense for the aging Mpofu. He
was past his prime at the age of ten and heading toward his twilight
years. Normally, a stronger male or coalition would have ousted him
from his cozy tenure. But with three strong sons bolstering the strength
of the family firm, he now had a new lease of life.[65] It makes evolutionary sense to team up with other close relatives. On average, sons
are as related to their father as to the half-brothers they'd usually team
up with. We named the three young males the Askaris—*askaris* being
young soldiers who would act as bodyguards to a chief.

The young Askari males were between three and four years old,
their manes and bodies just beginning to fill out to their future impressive size. The oldest of the three was a handsome cat with a perfectly
regal profile; the field team named him the Lion of Judah, or Judah for
short. He was missing the black tuft at the tip of his tail but made up for

this imperfection with an impressively dark mane even at his young age. We named the other two Jericho and Job, to round out the biblically baptized trio. Jericho was blond, and his mane remained light-colored throughout his life. Job was extremely shy and earned himself the nickname Scaredy Cat amongst the tourist guides. He would vanish into thick bush as soon as he saw a vehicle. Jericho and Judah were more self-assured.

At about the time that Mpofu and the Askaris were consolidating their tenure in the Kennedy pride, two young males were invading the study area from the south. Their names were Leander and Cecil. Nic and Jane first spotted these insurgents at Mangisihole Pan at the end of 2007. They were thereafter known as the Mangisihole coalition. The four-and-a-half-year-old Leander and Cecil steadily encroached on the territory of the Ngweshla pride, which had recently lost their pride males to trophy hunters. The Ngweshla pride occupied a prime territory in the eastern part of Hwange, in the area surrounding the water hole of the same name. Their range was rich in buffalo, giraffes, zebras, and kudu, thanks to the open grazing land extending eastward from Ngweshla. I'd known the pride since 2001, when I'd collared one of the five young females. The lionesses were called Posh, Baby, Scary, Sporty, and Ginger—named by a camp manager who must have been a fan of the '90s band the Spice Girls. We've followed the fortunes of this pride over several generations—keeping tabs on the daughters, granddaughters, and great-granddaughters of the original Spice Girls. They have always been a large pride; because of the abundant food in their territory, its lionesses almost always produce bountiful litters of cubs. There are frequently twenty or more lions in the Ngweshla territory. Such large prides and their productive territories are attractive prizes for male coalitions. The Ngweshla pride's males have had to fight hard to keep their tenure. The problem with owning something valuable—in this case a large pride and prime real estate—is that you attract envious neighbors.

By the end of 2008, Cecil and Leander had firmly established themselves in the Ngweshla range and killed the Ngweshla pride's nine cubs, fathered by the previous territorial male. Joanne, Jane, and I headed to the Ngweshla area to find and collar the new males. We found them each mating with one of the Ngweshla females close to the Makalolo Plains camp. Cecil was by far the more confident of the two. We were able to approach and dart him while his attention was distracted by the lionesses.

Cecil and his brother were a formidable coalition at the peak of their strength, but they could not hope to compete with Mpofu and the Askaris, their covetous neighbors to the north. Numbers count in battles between male coalitions, and two versus four is not good odds. Mpofu and the three Askaris began to make frequent forays into the Ngweshla territory to reconnoiter the enemy positions. Slowly and inexorably they mounted an invasion, spending more and more time in the heart of the Ngweshla range. Several running battles between the two coalitions ensued.

Five years later, in the shade of one of the giant ebony trees near the Ngweshla water hole, two big male lions—blond-maned Jericho and dark-maned Cecil—rubbed faces as affectionately as house cats. To place head and neck in such a vulnerable position, within reach of powerful forepaws and lethal canines, meant their trust in one another was absolute. Yet five years previously these two lions had been mortal enemies. Jericho, who had initiated the greeting ceremony, had been part of the coalition that had cornered and killed Leander, Cecil's brother. The bloody takeover had cost Cecil his coalition partner, his cubs, and control of the Ngweshla females. Though Jericho and his brothers had been victorious, Mpofu, their father, had been mortally injured in the bellowing, slashing melee at Three Pans, south of Makalolo.

The vicious battle of July 2009 saw Jericho and his brothers oust Cecil from the Ngweshla pride. Jericho, Judah, and Job—the Askari coalition—took over the prime Ngweshla range, killed the Mangisihole coalition's ten young cubs, and mated with the pride females. This should have been enough for the Askaris, but male lions are inveterate wanderers and are biologically hardwired to maximize mating opportunities. Once their cubs had been born to the Ngweshla females, the males spent more and more time away from the pride. Perhaps they were looking for some peace and quiet. More likely, they were looking for other prides with whom to consort. Their wandering led them to the park boundary. Soon a trophy hunter felled Job. Five months later, in October 2010, Judah went missing, almost certainly trophy-hunted. Once part of an invincible trio, Jericho was now on his own.

After the battle at Three Pans, the exiled Cecil moved eastward and took over another pride, the Backpans, who, in a familiar pattern, had recently lost its pride male to hunters. Cecil killed their cubs and set about making his own. The Backpans were another big pride; their range covered the spectacular Ilala palm–dotted grasslands in the east of Hwange. Wilderness Safaris leases this area as a photographic concession for several luxury lodges. Cecil and the Backpans pride were star attractions.

By the end of 2013, Jericho was nine and Cecil was ten. Both were past their prime. Being singleton pride males with tenure of large prides and rich ranges, they would inevitably be challenged by younger, hungrier males. Both were displaced from their separate prides by a pair of young males we called Bush and Bhubesi. In a peculiar symmetry, these males were the grown sons of the Dynamite males, the coalition that had displaced Mpofu from his original pride tenure eight years before. Cecil put up a significant fight. One evening in October 2013, we could hear the roaring and snarling as Cecil confronted the two interlopers in the scrubland west of Backpans. Two young males against a cat who, despite his age, was strong and wily must have been an extraordinary battle. But the odds were not in Cecil's favor.

Textbooks would predict that, expelled from their prides, Cecil and Jericho were at the end of their reproductive careers and possibly the

Cecil with the
Backpans pride.

end of their lives. But the social lives of lions are endlessly intricate. Who could have foreseen that, having both been usurped by the same coalition, they would nuzzle like males who had grown up together as cubs? Who would have anticipated that Cecil and Jericho would

forge an unlikely alliance and once again take control of the Ngweshla pride—a pride they'd both once held as partners in rival coalitions.

Male lions live tumultuous and complicated lives. The competition to mate and pass on their genes is ferocious. The price of a male's failure

Mpofu, after he was injured in a fight with Cecil and Leander.

is often death and the death of his cubs. Lions have evolved complex social strategies to limit this risk. Teaming up with brothers is the most evolutionarily advantageous strategy. But because forming bonds with other males is so critical to survival, joining forces with strangers or, rarely, a father-son alliance, are other options. This extraordinary behavioral flexibility allows males and their genes to survive the turmoil of lion society.

Contrary to what is generally thought, male lions are able to breed until they are eleven or twelve years old. In the last years of their lives, Mpofu, Jericho, and Cecil[66] all sired cubs. It is a fallacy that old males can be trophy-hunted with little disruption to lion society. In my experience it is extremely rare for a male lion to be without a pride. Deliberate removal of any animal from a population of social animals disrupts the web of social bonds and creates ripples of disturbance across a social system in which competitive relationships are often precariously balanced. The more I study lions, the more I come to realize how conservation

Jericho and Cecil.

managers underestimate the importance of this web of social bonds to the survival of individuals. Snipping through the strands of the web rends the fabric of lion society.

Mpofu's condition deteriorated to such an extent that we asked a veterinarian to have a look at him. The vet shook his head; there was no way he would recover. In November 2009 the area manager of Hwange National Park made the difficult decision that Mpofu should be euthanized. The old lion was spending almost all his time close to a picnic site where park visitors could stretch their legs or even camp for the night. An injured, hungry lion was too much of a risk. Although he was euthanized, he'd effectively lived out his life in the wild.

Jericho died of old age less than a kilometer from where Mpofu had died several years earlier. He'd once again lost tenure of the Ngweshla pride. In October 2015 Jane and I were radio-tracking him. The signal from his collar was booming through the receiver's small speaker; he had to be close by. I narrowed the location down to a clump of hook thorn acacia twenty meters away. Jane maneuvered the vehicle around

Burying Jericho.

the clump. At any moment I expected to get a glimpse of Jericho's distinctive blond mane in the shady interior of the island of bushes. It took me a moment to comprehend that the tufts of matted fur scattered at the edge of the clump were the remains of Jericho.[67] With a sinking feeling, I knew that the last of the great male lions in the story of two spectacular coalitions was dead. His body was positioned looking out across the open plain of the Masumamalisha Valley. We were sure he had died of natural causes, which was later confirmed by the veterinarian we called to perform a field postmortem. We buried the old warrior where we found him. We could not stand the thought that souvenir hunters might scavenge his remains.

# ELEVEN

# DEATH OF CECIL

The air bore the chill of the southern hemisphere winter, the Kalahari sand readily giving up the day's heat to a clear, star-peppered sky. Jackals yipped in the distance. Fiery-necked nightjars called to each other, their shrill onomatopoeic cry—"Good Lord deliver us"—was a plaintive supplication to the silent gods of the African wilderness. Otherwise the night was still. Most animals had moved out of the grassy valley to find warmer spots in the surrounding teak forest until the sun once again breathed heat into the land.

Cecil, the twelve-year-old male lion, padded along the dirt track with leisurely strides, soundless except for the crystal scratch of sand under his soup plate–sized feet. His coal-black mane proclaimed his status as the undisputed king of this part of the savannah. His pace was deliberate yet unhurried, calculated to eat up the miles, expending as little energy as possible as the huge cat patrolled his nocturnal kingdom. He paused only to scent-spray roadside bushes, maintaining his domain's signposts in a routine he had followed every night since he had become a territorial male nearly a decade before. He underlined his aromatic signature with vigorous scrapes of his hind paws.

The scent of a dead elephant drew the lion forward, enticing him

Cecil.

to what long experience had taught him was another free but odiferous meal. He had often fed on elephants. Sometimes he had hunted them; sometimes he fed on those that had died in the bush. Lions are as much scavengers as hunters. I have often seen them quite happily feed on putrid meat—wolfing down chunks of rotting flesh while blowfly maggots squirm like animated rice grains across their faces. The law of the

African savannah says, "Never waste free protein; starvation is only a few days away."

The carcass of the bull elephant had been in the sun for several days. Discarded remains of a trophy hunter's kill, stripped of ivory and skin and most of the edible meat. It was beginning to desiccate in the dry winter air, but enough flesh remained for scavengers to gnaw. A ripe smell permeated the bush for hundreds of meters, the stench as mouth-watering to a predator as Sunday lunch in the oven.

But there was something different about this carcass, something deliberate in its careful positioning in a clearing, something beyond this cat's experience of things to avoid. He could sense the presence of humans. No matter how quiet we think we are, how little scent we think we exude, animals pick up the tiniest cues. The rustle of clothing, the smell of toothpaste and deodorant, gun oil and plastic—they all stand out in a wild animal's sensory world like a snowflake in a coal mine. Humans did not worry him. He was used to their scent from years of living in a prime photographic safari concession in the park. But these were not the humans he knew. To minimize the scent and sound that would drift across the clearing, these humans were hiding in a tree platform downwind of the carcass. Crouched on a small platform was an American with a broad, white smile, a powerful compound bow, and a quiver full of lethally sharp arrows. He was flanked by a stocky Zimbabwean guide.

It would have been freezing cold to sit agonizingly still in the cramped hide. However, the hunters would have comforted themselves that the wait would not be long. This was an easy lion to hunt—a park lion, well-fed and habituated to people. Not like the mangy communal land lions that were so persecuted they vanished at the first hint of human presence. The locals boasted that you could drive right up to the cats around here. They would follow the trail left by the carcass the hunters had dragged close to the tree where the hunting blind had been constructed.

The big cat sniffed the clearing. The draw of the elephant meat overcame the lion's caution and he approached the carcass. He settled down to feed, tearing at the tough, dry meat with scissor-like teeth. He fed for a few minutes, oblivious to the hunter taking up the tension on his bow.

Cecil was the forty-second collared male study animal to be trophy-hunted since the lion project started in 1999. He was the sixteenth collared lion to be shot on that particular piece of land, Antoinette farm, a twenty-five-square-kilometer parcel bordering the national park. In some kind of awful symmetry, this was the same place where, fifteen years before, my first study animals, Stumpy Tail and Black Mane, had met their ends.

While this slaughter seems tragic, trophy hunting is a legally sanctioned use of wildlife in the hunting areas around Hwange. Trophy hunters do not break the law when they engage in permitted hunting safaris in this area. Hunting is an important part of the local wildlife economy. However, in 2015, the year Cecil was shot, the Gwaai Conservancy, the privately owned wildlife area adjacent to Hwange in which Antoinette farm is situated, did not have an officially approved annual hunting quota for lions.

Prior to 2015 the Gwaai Conservancy had been allocated a quota of between two and three lions. The conservancy's wildlife had been heavily depleted due to rampant poaching and uncontrolled hunting, and there was nothing like the wildlife that had been there in the conservancy's late-1990s heyday. Because there was little wildlife, there were also few lions. Invariably, lions hunted in the Gwaai were hunted directly adjacent to the national park boundary. These lions were almost always known animals from the study population in the park. By 2015 few ethical hunting guides would work in the Gwaai, so many of the hunts were farmed out to maverick Zimbabwean hunters or South African hunting companies. These fly-by-night outfitters had no stake in the long-term viability of populations and had little vested interest in looking after the wildlife resource that they were exploiting. Stories of hunting excesses and abuses were commonplace. There were stories of hunters using piles of watermelons and citrus fruit to lure elephants out of the park, and even rumors of occasional hunts taking place in the more remote parts of the national park.

In an effort to better manage lion trophy hunting, the National Parks, after initially banning lion hunting, had implemented a system to allocate lion quotas sustainably. This was based on an assessment of the ages of lions hunted in the area the previous year. The allowable quota was lowered if young animals were hunted and supplemented if only males past their prime were taken. In the National Parks' theory, hunting has little impact on lion populations if only old males are hunted.[68] In 2013 and 2014 most of the lions hunted in the Gwaai were subadults, and as a consequence of this poor record, the area lost its lion-hunting quota for the 2015 season. So the question in everyone's mind was: How could Cecil have been trophy-hunted on a Gwaai farm when there was no quota?

The circumstances surrounding this hunting incident are at best murky, and the facts were later widely misreported by the media. The following account is based on interviews with people involved in the hunt, statements made by those involved, and analysis of the location data collected via satellite from the GPS collar Cecil wore at the time he died.

We know that Cecil was shot by American bowhunter Walter Palmer late on July 1. At first, the hunters kept quiet about this hunt. They had every motive to be. The most salient reason for their reticence being that there was no officially approved lion-hunting quota. Research project staff only became aware that something was amiss six days after the incident. Project field assistant Brent Stapelkamp was routinely checking all the GPS downloads from the satellite collars we currently had fitted on the study lions. He noticed that Cecil's satellite collar had not transmitted any data since July 4. Initially he assumed that the collar had malfunctioned, although this seemed surprising given that the collar had recently been fitted and its batteries were new. Then, on July 7, project staff started to hear that a lion had been hunted in the Gwaai. In the small Hwange community, nothing of any consequence stays secret for long. Brent was sufficiently concerned to alert the National Parks management staff that an illegal hunt may have taken place. The park senior ecologist replied, "No legal hunt for a lion this year," and asked that the seemingly illegal hunt be reported to the National Parks wildlife officers at Hwange Main Camp.

Since there was no paperwork for a lion hunt in the areas concerned and no quota to hunt a lion, the senior wildlife officer ordered an investigation. National Parks requested that the project assist them by providing transport. Andrea Sibanda, one of the project field assistants, duly drove to the Antoinette area with a National Parks ranger to investigate the rumors. Andrea started his conservation career as an antipoaching ranger. His detective training in wildlife crime came in handy over the following days.

Andrea and the ranger's first port of call was the Hide, a photographic safari lodge nestled directly across the railway line from Antoinette. A friend of Andrea's had mentioned that, a few nights before, the hunting camp staff from Antoinette farm had visited the staff quarters of the camp. They were flush with money and looking to buy booze. On consuming the same, they had become talkative and were soon boasting about the huge lion that had been hunted a few days before. This successful hunt had resulted in their receiving a large tip from a very satisfied trophy hunter. The Hide camp staff suspected that the lion killed was most likely Cecil, one of the two magnificent male lions in which they had a proprietorial pride. Only one of the area's males, Jericho, had been seen since July 1, and he had spent several nights calling—in their opinion calling for his dead friend Cecil.

Armed with this information, the ranger and Andrea went across the railway to gather further evidence. The wily Andrea soon extracted the information out of the hunting camp staff working on Antoinette. The park ranger obtained signed statements from tracker Cornelius Ncube, who had assisted with the hunt, and camp skinner Ndabezinhle Ndebele, who had skinned the dead lion. These statements provided insight into how the hunt was arranged and conducted.

According to Cornelius, another hunting client had shot an elephant the previous week. The owner of Antoinette farm, Honest Ndlovu, had instructed the camp staff to keep an eye on the elephant carcass and to inform him if any lions came to feed. As it happened, two large males and a pride of females came to feed on the carcass the night after the hunt. This was duly reported to Ndlovu. On July 1, Cornelius was instructed to prepare for a lion hunt, which would be undertaken by a foreign hunting client, later identified as Dr. Walter Palmer, a dentist from Minnesota. Guiding Palmer would be Zimbabwean professional hunter Theo Bronk-horst and his son, Zane. The hunting party arrived at the Antoinette camp at midmorning and settled into its rustic accommodation. In the late afternoon, Bronkhorst took Cornelius to the still-reeking elephant carcass, which they moved, presumably by dragging it behind a Land Cruiser to a suitable location, approximately 300 meters away. Cornelius then assisted with the construction of a platform and hunting blind in a nearby tree overlooking the elephant carcass. Blind completed, Corne-lius was then driven back to camp. Bronkhorst and Palmer later returned to wait for a lion. In the early hours of the next morning Bronkhorst returned to the camp and woke up Cornelius, instructing him to come and assist them with a wounded lion. The professional hunter stated that they "had shot a lion with a bow and arrow and they were waiting for it to die." This is somewhat at odds with Bronkhorst's own account of the incident—as related to Peta Thornycroft, a reporter for Britain's *Telegraph*—in which he claimed he was unsure whether the lion had been hit by Palmer's arrow. Cornelius returned to the scene of the hunt with Bronkhorst and noted that in the darkness he could "hear [the lion] struggling to breathe."

It is clear that Cecil was at this stage mortally wounded and had not moved far from where he was shot. This is corroborated by the GPS data from Cecil's collar, which allows a forensic reconstruction of events. The collar sent a position from the hunt site at just before 9:00 p.m. By 11:00 p.m., the collar's position had moved eighty meters roughly southeast from the carcass. It therefore seems probable Cecil was shot at some point between 9:00 and 11:00 p.m. on July 1. Subsequent positions sent from Cecil's collar show he moved in a southeasterly direction until 7:00 a.m. on July 2. In around eight hours, the wounded animal had only moved 160 meters from the point at which he had been shot. Eventually, according to Cornelius, Bronkhorst advised Palmer to "finish the lion off." If Bronkhorst's later statements are accurate, the hunters went to administer a coup de grace at around 9:00 a.m.[69] Leaving Cornelius in the hunting blind, the pair went off in the vehicle to find the lion. According to the collar's GPS data, Cecil had by now moved a distance of around 350 meters from the point he was wounded. They killed him with a second arrow. Bronkhorst and Palmer returned to the hunting blind about forty-five minutes later with the dead lion loaded in the back of the hunting vehicle.

In media reports it was widely touted that Cecil suffered in agony for forty hours. This claim is inaccurate and exaggerated. It is unlikely he would have lived this long with such a severe thoracic injury. However, he most definitely did not die instantly and almost certainly suffered considerably. Judging from the events described by Cornelius and the data sent by the GPS collar, the injured lion was most likely killed around ten to twelve hours after being wounded.

The hunters then returned to the camp, and Cornelius and Ndabezinhle, the skinner, were instructed to skin the dead lion and begin preservation of the trophy. This involved removal and salting of the skin, which must be done promptly to avoid damage to the hide. Later the head would have been removed from the carcass, the tissue stripped off it. It would then be boiled and cleaned to the bone. Together, the skin and cleaned skull make up the "trophy" that a hunter would take home for display. Media reports made much of the seemingly barbaric "beheading" of the lion. That is, of course, effectively what happened, though perhaps in less sensational and bloodthirsty circumstances than most imagined. Cornelius and Ndabezinhle were ordered to leave the skinned carcass intact and load it onto the hunter's vehicle, along with the preserved skin. This was unusual, as the carcass, minus skin and head, has little value and is usually discarded in situ. Bronkhorst and Palmer then drove off, according to both Cornelius and Ndabezinhle, heading for Matetsi a few hours' drive away. It seems likely that, well aware there was no quota for a lion to be hunted on Antoinette farm, Bronkhorst was removing any evidence of the hunt. It is also probable that he was intending to report the lion as having been hunted in Matetsi Safari Area, or one of the other hunting areas to the northwest of Hwange, where there were lions on the hunting quota. This administrative sleight of hand is known as "quota swapping" and is unfortunately common in the hunting industry.[70]

There were other anomalies in the case that carry a heavy whiff of impropriety. Every trophy hunt undertaken in Zimbabwe is required by law to be accompanied by an official document, approved and stamped before the hunt by a parks officer. This document, known as form TR2, provides details of the hunt, including the hunter and client undertaking the hunt, its dates and duration, the location, and the species permitted to be hunted. The last, in theory, should be cross-referenced against documentation to show there is a valid hunting quota for the species in question in the hunting area. Without a valid TR2, trophies cannot be legally exported from the country. The reactions of the National Parks staff at Hwange Main Camp and their initiation of an investigation into the hunt suggests they were unaware of any approval. Bronkhorst clearly could not obtain a valid TR2 for a lion hunt in the Gwaai, as this area had not been allocated a lion-hunting quota in 2015. Instead, it seems he had a TR2 validated for a leopard hunt on Antoinette. It is not clear whether he in fact intended to hunt a leopard with his client or whether this was merely a subterfuge to disguise the intention to hunt a lion.

Additionally, the National Parks manager of the area had previously mandated that a ranger accompany trophy hunts for lions in the Gwaai area, to ensure all necessary regulations were followed. This local regulation was not adhered to.

Cecil marking his territory.

Another thread of evidence suggests that the hunters had every intention of concealing their activities. Cecil's satellite collar had functioned perfectly until 6:53 a.m. on July 4, two days after the lion was killed. Thereafter it ceased to send any further information and vanished without a trace. This seems an odd coincidence. Cornelius, the tracker, says that when he saw the dead lion in the back of the hunting vehicle, there was no collar. But he noticed that "the mane looked separated in the neck as if it had previously been caught in a wire snare." This suggests that Bronkhorst and Palmer removed the collar in the forty-five minutes between driving off in search of the wounded animal and returning with the lion's body in the vehicle. Neither Palmer nor Bronkhorst denied that the lion was collared, and both later expressed regret that they had killed a collared study animal. Bronkhorst claimed in media interviews that he had not known the lion was collared and would not have hunted it if he had. Palmer, once his involvement had been publicly revealed, stated that he "had no idea that the lion [he] took was a known, local favorite, was collared and part of a study until the end of the hunt." Bronkhorst admits that he "panicked" when he first saw the collar, which he then removed and hung in a tree close to where the lion had been killed. He later bemoaned the fact that he had not handed it in to the National Parks offices. Clearly this would have been the responsible and ethical course of action, though one that would have required an explanation of how his client had come to hunt a lion in an area with no hunting quota. From the GPS positions the collar transmitted via satellite over the next two days, one can see that, from the tree in which Bronkhorst abandoned it, the collar was moved several times. It sent positions from the main road to the hunting camp, and then from the boundary road of the farm, before being left on a neighboring property, possibly to implicate others in the killing of the lion. On July 4 the collar was once again moved, this time onto the railway line that separates the park from Antoinette. From here it sent its last position. Searches of this area for the collar proved fruitless. It seems likely that the collar was destroyed or perhaps buried deep enough that no radio signal could be received and no satellite communication was possible.

It is noteworthy that in hunts of collared lions, undertaken legally, the collars are invariably returned to the project research staff, or at least handed in to a National Parks office. When hunts are illegal or breach regulations, hunters with a bad conscience either destroy the collars or dump them somewhere to put any investigation off the scent. Bronkhorst claims that after hanging the satellite collar in the tree next to the dead lion, he never saw it again. This may be true, in that he did not himself move the collar. But clearly, once removed from the lion, it did not move by itself; someone involved in the hunt probably attempted to hide it and may have ultimately destroyed this critical piece of evidence.

Theo Bronkhorst is an experienced professional hunter who was not a newcomer to the area. It seems likely that he was well aware that any large male lions coming to the bait he had set up were highly likely to have come from the national park, little more than two kilometers away. The fact that the Hwange lion project collared many of the large males was not a secret. The hunting staff at Antoinette would almost certainly have been aware and would likely have seen Cecil on the property before, as it is an area he frequented regularly. However, it is also true that the collars are difficult to see when the lion has a large mane, as Cecil had. In fact, many tourists do not notice the collars at all. Perhaps Bronkhorst is sincere in claiming that he "could not have seen the collar at night. We would never shoot a collared animal." But I find it hard to believe that, once such a magnificent animal was in the hunter's sights, there could have been any other outcome. Certainly collars have never prevented hunters from shooting other study animals.

Walter Palmer allegedly paid $50,000 to shoot Cecil the lion with his hunting bow. He did not deny he had hunted the lion, nor that the hunt had taken place in a restricted area. He claimed to have relied upon his hunting guide, Bronkhorst, to obtain the necessary permissions,[71] which is not unreasonable given that he was paying a considerable sum for the professional hunter's services. In a public statement, Palmer maintained, "Again, I deeply regret that my pursuit of an activity I love and practice responsibly and legally resulted in the taking of this lion."[72] It is possible that Palmer was remorseful and—in hindsight, knowing the hunt was likely to have been illegal—regretted his involvement. Nevertheless, this was not the first time that Palmer was alleged to have been involved in an illegal hunting trip. Following the furor over the Cecil hunt, American media uncovered, through Freedom of Information requests, his participation in a similar incident nine years earlier.[73] In this incident, Palmer shot, again with a hunting bow, a large black bear in Wisconsin. Yes, he had a permit to hunt a bear, but he reportedly shot it forty miles from his permit's stipulated hunting area.[74] It is alleged that he subsequently offered substantial financial inducements to his hunting guides to lie about the location of the hunt.

The last time I photographed Cecil, May 2015.

Unfortunately for Palmer, the guides got cold feet and spilled the beans when the United States Fish and Wildlife Service investigated the incident. Palmer was charged with and, according to media accounts, found guilty of making false statements to a federal investigator, a felony. He received a fine of just less than $3,000 and a year's probation.[75] For a wealthy dentist, this effectively amounted to a slap on the wrist. He could potentially have faced five years in prison and a $250,000 fine. It is clear that Walter Palmer actively connived with his hunting guide to hunt a bear illegally. He then instigated, according to US Attorney John Vaudreuil, "a fairly aggressive cover-up"[76] and, according to media accounts, later turned on his erstwhile hunting companions, blaming them for the error. This mirrors at least some of the particulars in the Cecil hunting case. As with the bear hunt, the lion was killed in the wrong hunting area, with Palmer blaming his hunting guide for the error. We may never know if Palmer was in fact aware that the lion hunt was not permitted in the area. Again, so far, Palmer appears to have avoided any serious legal sanctions.

In both cases, Palmer appears to have gone to considerable lengths to obtain particularly large hunting trophies, exhibiting the obsession that many wealthy American hunters have for having their trophy records entered into the "trophy book" that hunting associations such as Safari Club International maintain for their members. Trophy books list hunting trophies by size, measured against a set of standardized criteria —noting the hunter's name and where the specimen was "collected." Horned animals are assessed on the size of their horns or antlers, predators on the size of their skulls, and pigs, hippos, and elephants on the length of their tusks. Hunters who submit particularly large "trophies" to a record book are feted by the hunting fraternity; to have one's hunting trophies recognized by one's hunting peers appears to be the aim of at least a subset of trophy hunters. This seems to many to be a distortion of what hunting is purported to be about. It takes no more skill or courage to kill a large rather than a small trophy animal. A buffalo bull carrying a record-breaking set of horns can kill you as easily as one with a mediocre spread. A mature lion with a huge skull succumbs to a high-velocity bullet with the same limb-entangled tumble as a subadult whose trophy measurement will never find its way into a record book. What it does take is money, and lots of it. Large animals are increasingly rare, as populations of most wild species decline and as hunters remove those species' largest members. It's expensive to travel to the limited remaining areas where large, old animals roam. It may take several trips to "collect" a suitably large specimen. Only the wealthy can play this game.

Cecil was a large animal, and one would have assumed Palmer was satisfied with his new trophy specimen. However, according to Bronkhorst, after the successful lion hunt, the American asked whether "[they] would find him an elephant larger than sixty-three pounds."[77] Elephant trophies are traditionally measured by the weight, in pounds, of the largest tusk. As Bronkhorst himself noted, this would require a prime-aged elephant of a size that is increasingly rare across Africa. Bronkhorst, perhaps regretfully, had to tell the avid hunter he was not able to find such a large elephant. It should also be noted that, as of April 2014, the United States Fish and Wildlife Service had banned all imports of elephant hunting trophies from Zimbabwe into the United States under the Endangered Species Act. This ban was still in place in 2015. If he'd succeeded in killing a giant pachyderm, it is unclear how Palmer, the collector of exceptional trophies, planned to get this one home.

What I find most difficult about the whole incident is the apparent callousness with which the hunters undertook this hunt. The lion was a commodity to be collected, "taken" in hunting parlance. Concern for the pain and suffering of the animal never seems to have been a particular consideration. I find the thought of killing any animal purely for sport or pleasure abhorrent, but if it has to happen, it must be done cleanly and without undue stress or suffering. Cecil suffered incredible cruelty for at least ten hours, severely wounded and slowly dying. Cornelius recalled hearing the animal "struggling to breathe." Clearly, though the wound was severe, the arrow had missed the vital organs or arteries that would have caused rapid blood loss and a relatively quick death. Certainly, the lion was so incapacitated it could only move 350 meters from the place where he was shot.

To kill a quarry animal quickly and efficiently is the hallmark of a good hunter. Yet Bronkhorst and Palmer had allowed an obviously stricken and mortally wounded animal to suffer for more than ten hours before attempting a final kill. Of course it was dark, and this may have made finding the animal more difficult. However, the lion did not move far and its distressing attempts to breathe could be clearly heard. It presumably could have been located in the relatively open country and been put out its misery. This the hunters did only the next morning between ten and twelve hours later. Perhaps part of the explanation is that Palmer was hoping to submit this obviously large trophy to a hunting record book as a bow-hunted specimen. This precluded the use of a firearm to dispatch the animal, as this would render the trophy ineligible as a bow-hunt record. The purists who oversee the "record book" stipulate that to qualify as a bow-hunted trophy, only a bow and arrow can be used.[78] One possibility, therefore, is that the hunters were content to leave the lion to die of its first catastrophic injury, allowing a bow-hunting trophy to be claimed with no further intervention. When the lion did not oblige them, they were, hours later, presented with the inconvenience of shooting it with another arrow. If this was the case, Cecil the lion died slowly and painfully to allow a hunter the ultimate vanity of claiming he had killed a huge lion with a bow and arrow.

I clearly recall the last time I saw Cecil. It was May 2015. Jane and I had been tracking him by following the signal from his collar. We eventually found him ambling along the dusty track between Ngweshla and the Masumamalisa Valley. We followed him a short distance before he flopped down on the road. From the scrub, spur fowl cackled their displeasure as he sat leisurely sniffing at the early evening breeze. We sat in the Land Cruiser a few meters away, taking photographs. He could not have been less concerned by our presence.

The research team was devastated by Cecil's unexpected death at the hands of callous men who cared little about the conservation of Hwange's lions. He was a spectacular and much-loved lion whose story we'd intimately followed over the eight years since I had first collared him in 2008. The hunting of Cecil the lion and the dubious circumstances surrounding the incident soon received global attention. It shone a spotlight on some of the questionable practices that are rife in the trophy-hunting industry. This was shocking to many, especially to people outside Africa who had no idea that trophy hunting of lions was legally sanctioned. It also led many to question whether, in modern society, it is acceptable to inflict any suffering on a highly sentient animal, let alone kill it, in the pursuit of sport or pleasure.

Cecil in 2011, an image widely used in the media after his death.

# TWELVE

# REMEMBER CECIL THE LION?

In the late evening of August 1, 2015, a lion, forty stories high, appeared in New York City. This was not a ghastly genetic experiment gone wrong, nor a leonine remake of *Godzilla*. The huge lion was an image projected onto the side of the Empire State Building in Midtown Manhattan. It was part of a show to promote the Discovery Channel's documentary film *Racing Extinction*. Throughout the evening, New Yorkers gaped at video images of rare and endangered species beamed onto the iconic building by powerful projectors. The lion was widely publicized in the media as being Cecil the lion.[79] The fact he was in the company of many other threatened species was in no small part due to the media attention given to the fate of lions after Cecil's death. It is pure coincidence that the show occurred almost exactly one month after Cecil the lion was shot by American trophy hunter Walter Palmer. But how was a lion from Hwange National Park in Zimbabwe elevated from an animal known only by local researchers and tourism operators to a global symbol in less than a month? How did a lion from a remote and obscure part of Africa grow in stature such that his image could comfortably fit alongside animal ambassadors for extinction—whales, tigers, and panda bears—with one of the world's most recognizable structures as a canvas?

At first, Cecil's death was felt only locally in the Hwange community. By July 7 or 8 rumors that Cecil had been shot started to gain currency. There was anger amongst the photographic safari guides who knew him well and for close to a decade had loved to take their guests to see the magnificent and highly habituated cat. He was "their lion," and they resented that a rich foreigner who knew nothing about his place in their lives had destroyed him. Everybody who worked in the tourism industry in Hwange knew just how easy it was to approach Cecil and how unchallenging it would have been to kill him. One of the safari camp managers posted a photograph on Facebook with the caption "RIP, Cecil." Reactions from safari guides and tourists who had known Cecil soon followed. Beks Ndlovu, a renowned Zimbabwean safari guide and CEO of a large photographic safari company, summed it up when he said, "In my personal capacity . . . I strongly object and vehemently disagree with the legalizing and practice of hunting lions in any given area. I will personally be encouraging Zimbabwe National Parks and engaging with Government Officials to stop the killing of lions and with immediate effect."[80] Meanwhile, in complete contrast to the reaction of photographic safari guides, professional hunters were overheard on their dedicated two-way radio channel congratulating Bronkhorst on the huge lion his client had "taken." Congratulatory comments on hunting websites went along the lines of: "Big old lion was killed by successful hunter in hunting concession south of Hwange. The lion was thirteen years of age and at the end of his natural life. What a story. Congratulations to Mr. Hunter."[81] This often deliberately defiant and sometimes goading braggadocio further incensed photographic safari guides and conservationists, provoking in turn strong and vociferous reactions about the practice of hunting lions on the boundaries of national parks.[82]

The global reaction to the hunting of Cecil may indeed have its roots in this more local conflict between photographic wildlife tourism and hunting safari operators in the Hwange area. Businesses deriving revenue from wildlife are effectively using shared resources. Lions, elephants, leopards, and other wild species hunted in the concessions close to the park boundary are the same animals that tourists view and photograph, animals that frequently become habituated to people through their contact with game-viewing safari guests. Safari guides and camp managers often operate in the same area over long periods and, as a result, feel an ownership of charismatic animals on which the quality and success of their guests' safari experience depends. This sense of ownership is particularly acute when the animal is individually identifiable, as Cecil was. On the other hand, at least on hunting concessions in the Gwaai Conservancy, professional hunters almost always buy hunting quotas from third parties. The hunters and their almost exclusively foreign clients have little vested interest in the hunting concession, often only visiting the area for the duration of the client's hunt. Neither professional hunter nor hunting client has any knowledge of the individual animals, and both have little awareness or respect for other users of the resource. In this sense it is as much a conflict over "rights" of locals versus "rights" of outsiders to use a resource. It is also about the markedly different way the resource is exploited. Photographic tourism requires a living animal; the resource can be reused as long as the animal lives. Known, long-lived, charismatic animals, resident within photographic concessions, have immense value under these circumstances. In contrast, hunting requires a dead animal; use is instantaneous and destructive—the ultimate consumerism. Individuality of particular animals has little value. The conflicting motivations of the users of the shared wildlife resource could not contrast more starkly.

The story of Cecil's death, as well as discussions about the ethics and legitimacy of hunting, started to appear on Internet discussion forums. Zimbabwe-based safari guides initiated them, but by the latter half of July 2015, a much wider public joined the conversation. It was clear that the hunting industry was becoming increasingly uneasy with the spotlight trained on it. News24[83] reported that the Zimbabwe

Cecil was a favorite amongst photographic safari tourists.

Professional Hunters and Guides Association, of which Bronkhorst was a member, announced that it was investigating the incident but had come to no conclusion. In the days following the exposure of the allegedly illegal hunt, National Parks' investigations branch deepened its probe. While offering few details, National Parks confirmed in a statement that "both the professional hunter and land owner had no permit or quota to justify the offtake of the lion and therefore are liable for the illegal hunt."[84] Both Theo Bronkhorst and Honest Ndlovu were arrested following the investigation and ordered to appear before the local magistrate on July 29 to face charges of carrying out an illegal hunt. The remains of Cecil, the skin and skull, were confiscated from Bronkhorst for use as evidence in the case.

While all this was going on, I was in Oxford, with a flight booked to return to Zimbabwe in mid-July. The field team had been keeping me updated about the situation on the ground. Journalist Adam Cruise, who was writing a blog for *National Geographic*, contacted me. His blog post and subsequent article[85] seemed to spur international media attention. Between July 23 and 26 several hundred articles were published and the story received wide coverage on news channels.[86] The level of attention paid to the death of a single lion seemed extraordinary, especially against a backdrop of a deteriorating security in the Middle East and Syria and a refugee crisis on Europe's doorstep. The sustained focus on an "animal story" surprised even seasoned conservationists. But the media storm was just getting started.

Although the local players had now been revealed, the identity of the foreign hunter remained in doubt. Some news outlets reported for close to ten days that he was Spanish. On July 27 Britain's *Telegraph* dropped a bomb. The killer of Cecil was not a Spaniard but an American dentist from Minnesota.[87] Walter Palmer's anonymity had been rudely stripped away. With an identifiable figure on which to coalesce the public's growing anger, the world's media went into a frenzy. Over the next twenty-four hours, the story grew exponentially—generating thousands of articles and tens of thousands of social media mentions. Celebrities started weighing in. The roster of movie stars, sports personalities, models, and entertainers commenting could easily have been the guest list for a glitzy Hollywood premiere.[88] Comedian Ricky Gervais posted a photograph of a lion on Twitter, re-tweeted more than 30,000 times, with the barely disguised rebuke: "I can't imagine anything more beautiful."[89] Sharon Osbourne was less subtle when she likened Palmer to Satan and suggested his dental patients should "take care."[90] Many social media posts were considerably less complimentary. The animal welfare organization PETA (People for the Ethical Treatment of Animals) called for Palmer to be "extradited, charged, and preferably hanged."[91] Actress Mia Farrow shared Palmer's business address with her 656,000 online followers, drawing criticism for appearing to incite irresponsible activism.[92] The story of Cecil's death rapidly went from newsworthy to viral. An impassioned monologue[93] delivered by popular American talk show host Jimmy Kimmel may have had a disproportionate effect in accelerating interest in Cecil. Kimmel's huge *Jimmy Kimmel Live!* audience watched as he deplored the unnecessary death of Cecil and pilloried Palmer. At the end of the piece, Kimmel's characteristically smooth delivery momentarily faltered as he wiped a tear from his eye and appealed to his viewers to show the world that "Americans are not like this jackhole." In an unprecedented move, he directed viewers to the Wildlife Conservation Research Unit's website, urging them to "make this into something positive." Over the next twenty-four hours, about 4.5 million people visited WildCRU.org, causing it and Oxford University's host Internet server to collapse—a first for one of the world's most prominent universities.

Both editorial and social media activity reached a crescendo the day after Kimmel's show and nearly a month after Cecil's death. As of July 29, 12,000 editorial articles were published in at least 125 languages.[94]

When you embark on a career in conservation biology, you never imagine that you will ever be close to the center of an event that attracts the attention of the international media. Certainly biologists receive little or no media training, so the experience can be overwhelming. I was now in Zimbabwe and had just returned to Harare after a few days' vacation in the Eastern Highlands with my young son. News stories about Cecil, my research project's study animal, seemed to be running on a loop on all the major news channels. Over the next few days my mobile phone hardly stopped ringing. I have no idea how journalists got ahold of my number. I found the experience deeply intrusive and disturbing. David Macdonald, the director of the research group, with the help of Oxford University's press office, bore the brunt of the media enquiries. At the height of the affair, he received 243 requests for media interviews in a single day and gave more than sixty separate interviews over the next fortnight. I also gave interviews from Harare. In one particularly surreal moment, I was waiting on cue for a Skype interview on BBC *Newsnight* with presenter Evan Davis.[95] I listened in as the preceding news piece on air discussed the unfolding Syrian refugee crisis. It seemed bizarre that a story about a lion I had first radio-collared six and a half years before was receiving the same level of interest as the calamitous global event affecting the lives of hundreds of thousands of people. In the hope of scooping a sensational story, the media descended on Hwange and besieged the project and National Parks staff. Some of the coverage was balanced but much was overdramatized. Some of the reporting was intentionally mischievous, and in one case, two news pieces generated completely contradictory headlines from exactly the same interview of project field assistant Brent Stapelkamp.

The extent to which the media attention was causing concern within Zimbabwe was underlined when I went to meet with the National Parks chief ecologist, Olivia Mfute. She told me that the issue had been discussed in a government cabinet meeting and that she was now having to field a lot of questions from anxious government ministers. In response to concerns, the government temporarily suspended hunting of lions and leopards in the hunting areas around Hwange.[96]

Radio-collaring Cecil,
October 2014.

Walter Palmer had returned to his home in Minnesota shortly after his lion hunt. After it emerged that he was the man who killed Cecil, he became, at least for a few weeks, possibly the most vilified man in America. Crowds of protestors picketed his home and dental practice and vandalized both premises. He allegedly received death threats. He went into hiding for several weeks and only returned to work in early September, once the media attention had started to die down.

Palmer was the perfect hate figure. He was a wealthy, white male who made no apology for his love of trophy hunting. This, along with evidence of previous hunting offenses[97] and alleged sexual harassment of a former dental assistant,[98] made him an easy target for public shaming. The media quickly unearthed photographs of him posing with the cadavers of animals he had bow-hunted. In one he poses bare-chested, the limp body of a dead leopard held tightly in his arms. As many people view this kind of posturing with uncomprehending revulsion, it is not hard to see how the public's negative attitude toward Palmer evolved.

Though plagued by the attention of the media and vociferous protests outside his home and office, he avoided any significant penalty. Initially there were calls from Zimbabwean government officials to have Palmer extradited to face charges of hunting illegally, although this was opposed by some, including Grace Mugabe, the politically powerful wife of former Zimbabwean President Robert Mugabe.[99] As the realities became clear, including the fact that Palmer appeared to have had the necessary paperwork to hunt legally in the country, charges against him were dropped in late 2015. Separately, the US Fish and Wildlife Service investigated his hunting activities in Zimbabwe for violation of the Lacey Act, a federal law used to prosecute wildlife trafficking offenses, including those taking place in foreign countries.[100] To date, no charges have been brought and a US Fish and Wildlife Service spokesperson stated in January 2018 that, "It's still under investigation. I can't comment further on a pending case."[101]

Theo Bronkhorst is a somewhat hapless figure. It is clear from statements issued by National Parks that he organized a lion hunt in an area where no officially sanctioned hunting quota had been approved. His activities can at best be described as shady. His role in the killing of Cecil has propelled him to international infamy. After his arrest, he was charged with "failure to prevent an illegal hunt." Those charges were dropped in November 2016, the court ruling that the allegations were "too vague to enable [Bronkhorst] to mount a proper defense."[102] However, to date, he has not been formally acquitted of hunting illegally and he did not escape punishment. His business was ruined, and he lost his hunter's license and his membership in the Zimbabwe Professional Hunters and Guides Association.

Soon after the Cecil incident, Bronkhorst was again in hot water, implicated in an attempt to smuggle sable antelope across the border into South Africa.[103] It is hard to be sympathetic toward the man who orchestrated the hunt for Cecil, yet he appears to have ended up as something of the fall guy. Bronkhorst has never denounced Palmer. While a more media-savvy individual would have sold his story to the ravenous media for a tidy sum, he never did. Bronkhorst's son, Zane, who is also a professional hunter and was present when Cecil was hunted, has been curiously airbrushed out of the story. Honest Ndlovu, the owner of the land on which Cecil was hunted and an active participant in facilitating the hunt, also appears to have escaped any significant consequences for his part in the affair.

The world reaction to Cecil's killing was astonishing and made headlines for several weeks. At the peak of the "Cecil affair," Cecil was mentioned in more than 94,000 articles between July and September. The response on social media was equally astounding, with 87,500 mentions of "Cecil the lion" on July 28 alone and more than 695,000 social media hits from July to September 2015. Few people with access to some form of news have not heard of Cecil the lion. During the height of the story, a friend from primary school with whom I'd lost touch for the better part of two decades, emailed me from Cambodia to say that Cecil was in the news there. This level of interest in a wildlife story, let alone the fate of an individual wild animal, was entirely unprecedented. Cecil, a lion from Hwange that few had ever seen, let alone known by name, became a household name. Something had bubbled up in the global consciousness, something that allowed an unknown African animal to eclipse the daily media diet of tragedy, war, and poverty. Conservationists are used to environmental stories being, if they are particularly lucky, relegated to page three or four or a few paragraphs in a special-interest section. Conservation issues hardly ever appear on the front page and never have the longevity of the Cecil story.

Why all of a sudden did the world appear to care? Most likely we will never know for sure. There are probably several aspects of the story that all contributed to creating something of a perfect storm.

First, at least in western Europe and North America, much of the population live urbanized lives divorced from the need to hunt wild animals for subsistence, or to protect livestock, crops, or their families.

An increasingly large and vocal subset of Western society views trophy hunting, the killing of an animal for sport or pleasure, to be morally despicable. In their eyes, killing Cecil was simply wrong. Almost all the reactions from celebrities and the general public were in this vein, with trophy hunting frequently described as "sick," "shameful," "disgusting," "shocking," and "disgraceful." Hunters themselves were reviled in similar terms. What is important here is not so much the reactions but that they were often the responses of mainstream opinion makers and therefore likely to represent, and perhaps even form, the views of the Western public at large. There are, of course, also many supporters of trophy hunting, but the reactions of proponents of hunting over the killing of Cecil were at best muted and were eclipsed by the outrage expressed by those who viewed hunting with distaste. The deep polarization of viewpoints between supporters and detractors of hunting are epitomized by the reactions of actor Sir Roger Moore, who played 007 spy James Bond, and Ted Nugent, rock musician, hunter, and gun activist. Moore characterized hunting as "a coward's pastime." [104] Nugent defended Palmer, stating that his critics were "stupid."[105]

The second significant factor that came out in the reaction to the hunting of Cecil was concern about suffering intentionally inflicted on animals. Many articles dwelt heavily on the report that Cecil, having been wounded by an arrow, suffered for "forty hours." While the timeframe was exaggerated in media reports, most people, even those who in principle support hunting, were rightly horrified at this obvious mistreatment of an animal.

Society puts significant emphasis on the question of culpability. To many the fact that these men went out with the intention of killing a lion purely for pleasure and to collect its head and skin puts it beyond the pale. The motivation is clearly not one of self-protection or protection of livelihood—the justification that might apply to a rural African killing a lion. Media reports made much of the "beheading" of Cecil's body, the ultimate act of culpability, referring in intentionally sensational and loaded terms to the collecting of the lion's skull as a trophy of the hunt. "Collection" of head, horns, antlers, and skin is accepted practice in trophy hunting, but many saw the mutilation of the dead animal as barbarous. We are similarly appalled by biologist Dian Fossey's heartbreaking account of finding the body of Digit, the silverback mountain gorilla with whom she had formed a bond. His head and hands had been hacked off by poachers, for sale to tourists as gruesome curios. In a fusion of anger and grief, we are outraged that his body has been mutilated. Who would do such a thing, and who in the world buys the severed hands of a gorilla as a holiday memento?

A third factor may also have been critical. This lion had a name. In a Tweet replete with irony, Louis Theroux advised his Twitter followers that "a good rule of thumb for hunters would be not to shoot any animal with a name. It's asking for trouble." Theroux identified the fact that people are much more likely to empathize with a named, rather than generic, individual. It was not just any lion; it was Cecil the lion. We name things to which we attach emotional importance and with which we identify. By naming individual beings, we are able to invest in them our values, preconceptions, and emotions and to imbue them with feelings of their own. Perhaps we go further in assigning rights to such creatures—the right to fair and ethical treatment, and even the right to life. Because Cecil was referred to by a name, he was transformed in our minds from beast to being. Mistreatment of a being, and particularly a charismatic being, elicited feelings of anger or revulsion.

Scientists try hard not to anthropomorphize their subjects—to avoid the subjectivity of interpreting animal behavior in terms of human values and motivations. Nevertheless, highly evolved animals, particularly mammals and birds, show behavior and have cognitive function not very different from our own. They experience fear, recognize friends and enemies, care for their offspring, make sacrifices for their social group and close kin, and form and maintain strong social bonds. Anyone who has spent any time closely observing animals, whether wild or domestic, will

Cecil.

tell you they have "personalities"—they react to a set of circumstances in a particular and predictable way that varies considerably between individuals. They are far from the preprogrammed automatons that evolutionary purists would have us believe.

Another aspect to this is the extent to which a well-recognized name can come to encapsulate a story. The name Cecil, or derivatives of it, is reasonably common in both the Anglophone and Francophone spheres. This made it memorable, much more so than an indecipherable African word or Cecil's scientific code of MAGM1. The fact that readers and viewers did not need to stumble over an unfamiliar or unpronounceable word made for slick headlines and easily tagged stories. By the end of the first week of the coverage, everyone knew who Cecil was and what had happened to him; there was no need for further explanation. This remains the case. So when, almost exactly two years later, headlines declared that Cecil's son Xanda had been shot by hunters,[106] few people needed to ask "Who is Cecil?" It was an immediately recognizable and digestible headline in which the name "Cecil" told the story.

However, the lion being named is not the only reason that the Cecil incident became so newsworthy. Recent deaths of other named large predators have caused little stir. Up until Cecil's death, hunters had shot forty-one named lions radio-collared by our project. None made the news. In the same year that Cecil died, two other large, charismatic predators were illegally killed: Legolas, a cheetah, and Scarface, a grizzly bear. Like Cecil, they were part of long-term scientific studies. Both were killed in seemingly murky circumstances. Researchers found Legolas on the side of the road, a spent shotgun cartridge nearby.[107] He'd clearly been shot by a passing motorist and left on the roadside, radio collar undamaged. This happened in a part of Botswana where farmers regularly kill cheetahs in defense of domestic livestock, which may have been the motivation for this seemingly senseless slaughter. Twenty-year-old Scarface had been studied by biologists in Yellowstone National Park. He was shot in November 2015 outside Yellowstone, probably by a hunter, though not one looking for a trophy, as his carcass was left untouched. There was some suggestion that Scarface may have been killed in self-defense, but this was never conclusively proved.[108] Like Cecil, he was a long-term study animal, having been radio-collared seventeen times; he was also frequently photographed by visitors to Yellowstone National Park. Both incidents received only limited international media exposure.

Why did these two news stories about well-studied, charismatic large predators fail to go "viral" in the same way the Cecil story did? There are several other aspects of the Cecil story that set it apart from those of Scarface and Legolas. In the Cecil case, the perpetrators of the killing were identified and were alleged to have acted illegally. In the case of Legolas and Scarface, their killers remained anonymous and the motivations and circumstances surrounding the killings ambiguous. Finally, one reason the cheetah and bear were ignored but the lion made headline news is that people and society value some species over others. Imagery of lions is omnipresent in our culture, from the nursery to heraldic imagery in public life. If I browse through my children's bookshelf, images of lions are everywhere. Bears are too but mostly of the cuddly teddy bear variety. Cheetahs hardly appear at all. Studies of animal charisma consistently show people far prefer large over small species, predators over herbivores, animals perceived to be rare, and species with forward-facing eyes.[109] Lions ranked third after tigers and African elephants, suggesting they are highly recognizable species that are greatly valued by society.

Caution is needed in interpreting the scope of the global media interest. We live in a world where the important issue of whether a dress appears blue and black or white and gold has the capacity to hold the world's open-mouthed attention.[110] Journalist Franklin Foer[111] takes a cynical but plausible view. He suggests that the seemingly endless generation of media stories about Cecil may, in part, be a function of the creation of "clickable" content by online media publications in order to compete for digital readership and the data-driven profit this brings. The generation of spurious and sensational articles such as "Cecil the Lion Shares Message of Hope to Animal Psychic's Facebook Page"[112] may just be a symptom of digital media interests surfing the wave of a viral trend in an attention-deficit world.

To echo Jimmy Kimmel, did anything positive come out of it?

Was the death of Cecil a watershed moment, and what does this mean for lion conservation? The killing of Cecil was arguably a disruptive event that shone a spotlight on the plight of African lions and the threatened ecosystems they rely on. Many people had little idea that there are fewer lions in Africa than there are rhinoceroses or elephants and that lion populations are declining rapidly in many parts of the continent. It also revealed some of the deplorable practices that are carried out in the name of conservation by some trophy hunters. People were astounded to discover that a significant proportion of wildlife habitat in Africa is protected and managed as hunting areas. Most of all, the incident started a conversation about how these delicate and imperiled African ecosystems can be protected for the future. It is also clear that moral outrage is not going to be enough, nor are the solutions likely to be simple or easy to implement. The future of African lions is inextricably intertwined with geopolitics, policy, historical wildlife-management practices, and deepening poverty on the African continent.

# THIRTEEN

# HUNTING FOR A SOLUTION

In November 2017, to the astonishment of many, US President Donald Trump reversed, pending further review, the recent decision of his Department of Interior to reinstate importation of big game trophies to the United States.[113] In a tweet he stated he would be "very hard pressed to change [his] mind that this horror show in any way helps conservation of elephants or any other animal." This statement by a conservative president left political commentators flummoxed. Though Trump himself has indicated his personal ambivalence to hunting, his sons are enthusiastic hunters and have drawn fire in the media by posing with animals they've killed on hunting trips to Africa.[114] The Republican Party, of which Trump is the leader, is the spiritual home of pro-hunting and pro-gun lobby groups. Differences within the Trump family and within the conservative establishment over the issue of trophy hunting may mirror similar schisms within society and signal a growing distaste for trophy hunting. Recent opinion polls suggest that 69 percent of Americans oppose trophy hunting and 78 percent oppose imports to the United States of trophies of hunted lions or elephants.[115] Given that a high proportion, in some cases up to 80 percent, of foreign trophy hunters visiting Africa are American,[116] policies regulating importation of hunting trophies have significant implications for conservation policy in Africa, where hunting has historically played a prominent role.

The shifting sands of the American political landscape may shape the way its citizens engage with nature in Africa, but the future of lions hinges on how they are valued by Africans, who must live alongside them and whose governments have to bear the costs of their conservation. These two realities are not as disparate as they may at first appear. While lions are culturally important in Africa, conservation is not high on the agenda in countries where food security, access to health care, and education are limited. African states lack the tax base and revenues to subsidize conservation, in contrast to Europe and North America, where taxpayers cover the cost of conservation and protected area management. Africa's leaders, like politicians everywhere, need to get elected, and few African leaders run for office on an environmental platform. Putting the welfare of animals before the welfare of people is never going to be popular. That is why conservationists have often sought to place a value on wildlife—to give people and governments a monetary incentive for sparing nature.

This is where trophy hunting comes in. The wealthy foreigner goes "on safari" and pays astronomical amounts to kill a wild animal that might otherwise be viewed by locals as worthless or even pestilential. Africans now have a motivation to keep otherwise unwanted wildlife around. That's the theory. It's a theory that has been around a while. In fact, African conservation has its roots in the policies of colonial administrations of the late nineteenth and early twentieth centuries.[117] Upper-class sportsmen, horrified at the wanton slaughter of wildlife by colonial settlers, formed the Society for the Preservation of the Wild Fauna of the Empire. Its members included Theodore Roosevelt, Lord Kitchener, and many other luminaries of the time. The society lobbied the British colonial administration to set aside "game reserves" where hunting could be regulated. Describing wild animals as "game"—quarry to be killed for sport or amusement—speaks to the attitudes and ethos of the time. Critics pilloried the society's members, many of whom had been avid hunters, as "penitent butchers." Nevertheless, many of the reserves set up in Africa at this time persist as modern-day national parks and safari areas. The original game laws governing these areas often form the basis of current conservation legislation.[118]

Outright protectionism has never been at the core of African conservation. As a result, about half of the land set aside for conservation on the continent, an area of approximately 1.2 million square kilometers[119] (twice the size of Texas and five times the size of the United Kingdom) is managed and protected expressly for trophy hunting. This comes as a surprise to many people unfamiliar with African conservation. It is certainly arguable that if this land had not been protected for hunting, it might have been settled and converted to farmland. If these areas were no longer protected for hunting, it is unlikely African governments would have the political will to convert these areas into national parks, or the funding to support such a move. This is really at the core of the conundrum surrounding the issue of trophy hunting. No matter how unpalatable hunting is to many, the simple fact is that hunting areas constitute close to half of all protected land in Africa. Loss of that wildlife habitat would be catastrophic, and settlement of this land by people is likely to be ecologically and politically irreversible.

In Africa, a well-worn mantra holds, "If it pays, it stays"—meaning that if conservation of wildlife is financially viable, protecting it will continue to be publicly defensible. So does hunting pay? Fifty thousand dollars is a whole lot of money. With this amount of cash in their metaphorical wallets, most people might put down a deposit on a family home, buy a dream motorcar, or kick-start a child's college fund. It is also the sum media articles claimed Walter Palmer paid to hunt and kill Cecil.[120] This is about the average fee charged for a lion hunt in southern Africa, though lion hunts in prime wildlife areas can sell for more than $100,000.[121]

The future of lions may depend on how they are viewed by Africans.

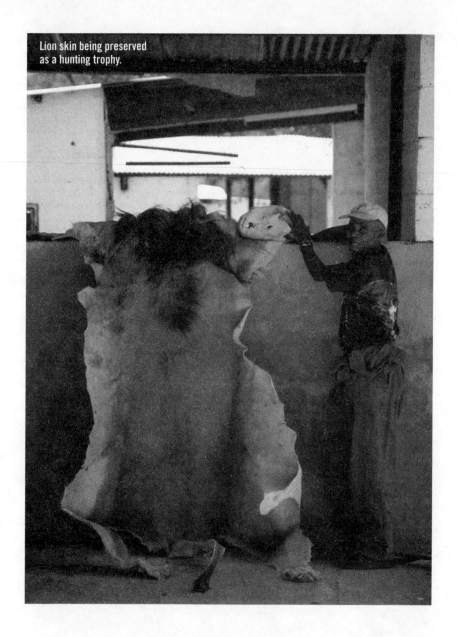

Lion skin being preserved as a hunting trophy.

The revenue earned by trophy hunting—and its potential to pay for conservation and to benefit poor local people—is one of the most prominent arguments used by hunters to justify their sport. On the face of it, this does not seem unreasonable. However, the argument is worth examining in a little more detail. First, does the revenue from hunting cover the costs of conservation? In Africa, on average, the annual cost of conservation (such as employment of park rangers, maintenance of infrastructure, and protection from poachers) comes in at around $500 per square kilometer.[122] This is actually quite modest compared to what is spent on reserves in North America, where conservation expenditure sits at around $2,500 per square kilometer.[123] According to conservation biologist Peter Lindsey, revenues from hunting concessions amount to around $400 per square kilometer per year.[124] Deduct hunting concession and trophy fees, then subtract operating costs and profits, and you discover that hunting revenue does not come close to covering the actual costs of conservation.

Let's look at this in a lion-centric way. Craig Packer calculated that it costs a minimum of $2,000 per square kilometer per year to protect a lion population at half its potential size in an unfenced reserve.[125] To extrapolate what it might have cost to protect Cecil from the time he was born to the time he died at the age of twelve, you might consider that he occupied around 500 square kilometers, the average range of a male lion in Hwange. It would arguably have cost $1 million a year to protect Cecil and the ecosystem that supported his existence. Over a twelve-year period, this would have amounted to $12 million dollars. Of course most African countries spend far less than $2,000 per square kilometer on park management. Hwange's annual conservation budget is around $276 per square kilometer. Even at this bargain-basement budget, it cost just more than $1.5 million to protect Cecil until he was twelve years old. A one-off fee of $50,000 to kill him did not remotely offset this cost. Nor did the park, whose budget paid for his protection, benefit from this revenue. In reality, hunting greatly undervalues African wildlife. That is not to say that people do not become rich through hunting. They do. But little of the financial gain filters down to covering costs of conserving wildlife.

So if hunting revenues do not, at least most of the time, cover conservation costs, do they at least benefit rural Africans, whose tolerance of wild animals is critical to the animals' survival? The answer depends on whom you ask. Some hunting operators do make a big effort to support local communities. However, I've seen almost no benefit to villagers around Hwange National Park—from hunting or, indeed, from any kind of conservation revenue. In our survey of 624 households in three communities close to the park, only 18 percent of the respondents said they gained any benefit from trophy hunting taking place on community land. Of those who said they benefitted, only a quarter gained any kind of direct financial benefit—on average relatively modest one-off payments averaging $40.[126] Given that the definition of local poverty is living on less than one dollar a day, trophy hunting is unlikely to raise living standards significantly. The promise of improved livelihoods through revenues derived from trophy hunting does not appear to be fulfilled, nor is hunting incentivizing people to tolerate wild animals.

So if the commoditization of wild animals by the hunting industry doesn't pay for conservation, the refrain of "If it pays, it stays" starts to sound worrisomely hollow and not dissimilar to a protection racket—offering the unpalatable certainty of poorly managed hunting against the perceived alternative that wildlife would disappear if hunting were ended.

Opponents of hunting often argue that trophy hunting can be replaced by photographic tourism. It is true that, like trophy hunting, photo tourism does generate significant revenue. Tourism in Botswana is expected to generate $210 million in 2017.[127] It is tricky to estimate how much wild animals are worth to the tourism industry since their value is woven into revenues from local transport, accommodation, meals, and staff earnings. Some estimates suggest that a single male lion could be worth as much as $100,000 per year, based on the time tourists spent viewing the animals.[128] However, this form of tourism is only viable in prime areas with extraordinary landscapes, such as the Okavango Delta in Botswana or the Serengeti in Tanzania. There is also a limited market for photographic tourism, with much of it focused on well-known destinations. A lot of the African continent is inaccessible—hot, dry, dusty, and worryingly insecure. Unless you love Africa, these areas can seem a little bit bleak, even frightening. What holiday-maker will pay to spend days on end searching for that elusive lion or elephant while roaming miles of broken scrub—with the added spice of a possible encounter with AK-47–toting insurgents? Photo-tourism has great potential to protect prime wildlife areas in politically stable regions. But is likely not a universal substitute for trophy hunting, particularly in areas that your average vacationer does not want to visit.

Growing up in Africa, I learned that assigning monetary value to wild animals and using them sustainably ensured their continued survival. By harvesting crocodile eggs in the wild, my father and his colleagues helped to set up a crocodile farming industry that brought crocodiles back from the brink of extinction, having been previously decimated by commercial croc hunters. Well-managed and well-regulated hunting areas have similarly benefited wildlife populations. But many hunting areas are poorly managed. I have grown skeptical that hunting can remain a sustainable activity, particularly with the poor governance prevalent across Africa. I worry that as overhunting and mismanagement deplete the "game" in hunting areas, and these areas become less commercially viable, they will be abandoned as wildlife preserves and surrendered to agriculture. This has already started to happen.

I've spent much of my career trying to understand the impact of trophy hunting on lions. I find it hard to grasp what motivates a sport hunter to kill a lion, an animal that is so much more beautiful alive than dead. An animal that poses no threat to the hunter, his family, or his livelihood—an animal that will not be eaten or used for anything other than a trophy for display. It seems obscene that an animal should die just to bolster the ego of a rich Westerner who wishes to adorn his home with a woodenly taxidermied replica of a creature that was once lithe and vibrant. Lion hunting is not particularly challenging or dangerous. Big cats are commonly shot by luring them with bait

and shooting them while they feed.[129] I've darted enough lions to say with some confidence that a feeding lion is relatively easy to approach and, when hungry, pretty much oblivious to any danger. A hunting acquaintance once admitted that much of a hunting safari is "smoke and mirrors" theater created for the benefit of a naive but wealthy foreign client. You have to wonder if hunting stories told in the man cave with a glass of whiskey bear much relation to the actual event.

Although I do not personally hunt, I'm less opposed to subsistence hunting. Hunting a deer on license in the local woods and stocking the freezer with venison to feed a family seems, morally, little different from buying a packaged steak at the supermarket. It is also environmentally more sustainable, since livestock farming contributes to deforestation, land conversion, and greenhouse emissions. A subsistence hunter is a far cry from a trophy hunter who spends vast amounts of money to travel to a foreign country for no other reason than to collect an animal's head and pelt to display in his billiards parlor.

In our study of lion population biology in Hwange National Park, trophy hunting had the single-most-significant effect, with levels of hunting mortality exceeding deaths of lions in conflict with people or killed in wire snares set by poachers. It far outstrips natural levels of mortality. But this is not the whole story.

I have spent considerable time over the last two decades watching lions. They never seem more content than when they are with other lions. They lie together at rest, draping their paws over snoring neighbors. When they wake or return to the pride after time away, they greet each other with obvious affection, rubbing faces and rolling on the ground together, uttering growls and grumbles of pure feline pleasure. None of this is surprising because, for lions, the social bonds within the group are critical for survival. These animals are cooperative in everything they do, suffer visible distress at the disappearance of a group member, and defend each other and their cubs to the death. Deliberately killing key individuals within the intensely social lion prides leads to a cascade of turmoil and disturbance that we are only beginning to understand.

In March 2016 tragedy befell the Ngamo pride. The pride had been out of the park in the Tsholotsho Communal Land for several days and had been killing domestic livestock. The Lion Guardians team had tried in vain to chase the lionesses and their cubs back into the park. They wouldn't budge. Back in the park, two male lions, strangers to the pride, were waiting. They had recently ousted Cecil's son, Xanda, the former territorial male and the father of the pride's cubs. His defeat had left the pride and his cubs undefended. Pride takeovers and infanticide by new territorial males are a natural part of lion behavior. But in this case, the ousting of Xanda had been precipitated by a trophy hunter killing Xanda's brother, Sinangeni. The loss of his coalition partner had made Xanda's tenure of the Ngamo pride precarious and vulnerable to the two invading males. The new males, Nqwele and Butch, were intent on killing the cubs sired by their former rival. The Ngamo lionesses would do anything they could to protect their cubs from the new males. In their desperation, the lions took refuge outside the park, ironically in a place where the risk of death was even higher. On March 5 the lionesses killed two donkeys. Angry villagers demanded something be done. That evening the district Problem Animal Control officer responded by shooting Inkosikasi, the pride's matriarch. The next morning, he shot her sister. These two females were the core of the pride. Both had swollen teats, indicating that they had been nursing cubs that would have been hidden in a thicket nearby while the females hunted. With their mothers dead, the cubs almost certainly starved to death or were dispatched by hyenas or stray dogs. Still the remnants of the pride would not return to the park. Over the next few weeks the entire Ngamo pride—three lionesses, a young male, and eight cubs—was wiped out.

The killing of one pride male by a trophy hunter had destabilized the precariously balanced structure of local lion society and led to the deaths of the rest of a pride. We've seen this pattern of disruption repeated almost every time a male lion is trophy-hunted in this part of the park.

Our humanity is defined by how we behave toward other sentient animals. Inhumane treatment of animals has become socially unacceptable. In many countries, laws now enforce the humane farming of domestic species. The use of animals in pharmaceutical and cosmetic testing has also become heavily regulated. This trend has, quite rightly, come to permeate the medical and behavioral sciences. To undertake behavioral studies on lions in Africa, I must present evidence to a university ethics committee that my work is necessary and that I have taken steps to ensure the welfare of study animals—including limiting any unnecessary stress or discomfort caused by study protocols.

Because of this recognition of a need to treat animals humanely, there is a significant chance that the days of African trophy hunting—at least the hunting of social species such as lions and elephants—are numbered. As evidenced by the outcry over the hunting of Cecil, much of the public no longer views hunting as an acceptable leisure activity. There is precedence for this shift in perception. Commercial hunting of whales in international waters was banned by the International Whaling Commission in 1986, partly because whale populations had declined drastically due to overexploitation by whaling nations[130] but also because public opposition to the killing of highly sentient mammals had reached a crescendo. Similar activism virtually closed the commercial fur trade and the unregulated trade in spotted cat skins and arguably gave rise to the formation of the Convention on International Trade in Endangered Species of Wild Fauna and Flora (CITES).

Concern about the inhumane treatment of animals has triggered bans of many traditional blood sports across Europe. There are strong parallels between the opposition to the hunting of foxes and the hunting of charismatic African species. Fox hunting became a central political issue in the United Kingdom in the 1990s. By 2004 Parliament had outlawed the centuries-old tradition of hunting with hounds.[131] Hunting foxes with hounds has virtually no demographic effect on fox populations. Indeed, it had demonstrable benefits to the conservation of rural biodiversity through the protection of wildlife habitat by hunt enthusiasts. Fox hunting was banned almost entirely on welfare grounds; the largely urban electorate viewed the killing of foxes for sport with deep distaste. There were also political overtones to this ban. A well-educated middle class saw fox hunting as one of the last bastions of privilege; perhaps metaphorically pulling the scarlet-coated aristocrats off their horses was in part a backlash against cultural elitism. These are almost precisely the prevalent public reactions to the killing of Cecil. As with fox hunting, African safari hunting is viewed as a pastime of a wealthy elite. People of more modest means are apt to be revolted by the killing of wild animals purely for sport.

There is another moral dimension to African trophy hunting. Regulated sport hunting by citizens is well developed in the West—that is, North America and Europe. It does not exist to the same extent in Africa. The majority of Africans do not hunt for pleasure or sport but rather to feed their families or to protect their livelihoods. When they do so in protected areas, they are called poachers. Wealthy Westerners who pay large sums to hunt the same animals, sometimes in dubious circumstances, are called "sportsmen." The former face legal penalties, including lengthy prison terms; the latter receive the accolades of their peers. There is an uncomfortable juxtaposition between the villager led off in handcuffs for killing a livestock-raiding lion and the Western hunter posing with the artfully arranged corpse of the collared lion he has just illegally hunted. There are starkly different rules for the elite and the poor. I wonder how long this will be acceptable to Africans.

The Cecil incident is a watershed moment in the history of hunting. Prior to Cecil's death, many people were unaware that it was even legal to hunt lions. Even those who were aware learned about the irregularities and mismanagement that beset the hunting industry. The level of public outrage exceeded anything generated by the sustained campaign to end fox hunting. The media has a notoriously short attention span. Nevertheless, even several years later, the Cecil story grabs headlines. If supporters of hunting hoped that the incident would be forgotten, they are sorely disappointed. Thanks to the actions of Bronkhorst and Palmer, the hunting industry is under pressure as never before. I was recently in the international departure lounge of Johannesburg Airport catching a connecting flight from Zimbabwe. To kill time, I browsed through the magazine section in one of the airport bookstores. Leafing through the hunting magazines,[132] I came across several articles specifically related to Cecil and the 2017 killing of Xanda, Cecil's son. The articles were defensive in tone and gave the clear impression that hunters remain worried about the future of lion trophy hunting.

Professional Hunter Ivan Carter articulated the industry's concern in an *African Hunter* article titled "The Hunters' Image."[133] Calling the Cecil incident the industry's "Twin Towers event," Carter recognized the need for hunting to be more explicitly about conservation and much less about ego. However, the public's negative perception of trophy hunting will not be easy to dislodge, even if the hunting industry belatedly decides to reform. I believe an increasingly large segment of society will continue to see the hunting of lions for sport as a distasteful anachronism, a throwback to the colonial era when upper-class explorers posed with the carcasses of the exotic animals they had "collected" while on safari.

There are already restrictions being placed. Botswana, a country in Africa with one of the best track records in conservation, banned all trophy hunting in the country in 2014. There is a good chance that in Europe and North America public opposition to hunting will sway conservation policies, including tighter regulations on trophy imports. Since the majority of trophy hunters also come from these regions, this could have a significant impact on the hunting industry. The US House of Representatives passed legislation to control trade in endangered or threatened wildlife, naming it, in honor of Cecil, the Conserving Ecosystems by Ceasing the Importation of Large (CECIL) Animal Trophies Act.[134] The issue of lion trophy hunting has been discussed in Parliament in the United Kingdom and the European Union. Both Australia[135] and France[136] banned imports of lion trophies in 2015. Many major airlines, including Delta, United, and American, have banned carriage of hunting trophies, in direct response to the public outcry over the killing of Cecil.[137] The European Union and the United States have reviewed their policies on importation of lion trophies, with the latter temporarily suspending imports pending reviews of exporting countries' conservation policies. President Trump's recent stance on trophy hunting may be a reasonably accurate barometer of public sentiment and could signal further restrictions of imports of trophies from hunted endangered species to the United States.

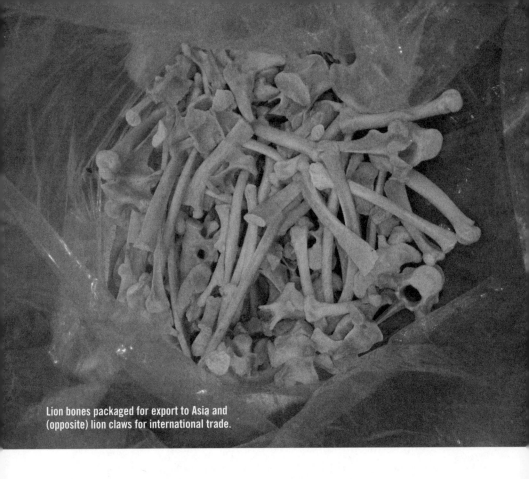

Lion bones packaged for export to Asia and (opposite) lion claws for international trade.

Trump's disquiet over hunting is echoed by former director of the US Fish and Wildlife Service, Dan Ashe, who is skeptical that trophy hunting is a useful tool in African conservation. Ashe, who describes himself as a "lifelong and proud" hunter, articulates the moral inconsistency that comes with outlawing hunting in the United States of native species listed as endangered yet allowing US citizens to import trophies from similarly imperiled species killed elsewhere. He states: "If elephants were native to the United States, and endangered or threatened, they would not be hunted. And neither would lions, rhinos, or leopards."[138]

There is a danger, though, that the trophy-hunting issue, while important, may distract policy makers from more critical threats to lions. Banning lion trophy hunting will not halt the decline of Africa's lion

populations. Loss of habitat and conflict with livestock owners are the most important causes of lion population decline. The threat of habitat loss may actually be exacerbated if hunting areas are no longer protected.

Another emerging threat may soon eclipse that of trophy hunting—namely, the illegal trade in lion body parts for use in traditional Asian medicines. The same trade decimated tiger populations. Possibly because the use of tiger parts is now strictly regulated, the demand for lion bones, teeth, and claws—perhaps as a substitute—appears to be growing. A popular use of big cat skeletons is for the production of lion and tiger bone wine. The "restorative tonic" is made by suspending a big cat's carcass in a vat of alcohol. Since 2008, there has been a significant increase in exportation of lion bones from South Africa, with up to seventy tons of lion bone exported mostly to Laos, Vietnam, and Thailand.[139] This trade is legal, and the bones are derived from farmed or captive lions or the discarded remains of trophy-hunted animals. Although CITES imposed restrictions on trade in lion bones in 2016, the demand is undiminished. With limitations placed on legal supply of bones, traders may already be illegally sourcing bone from wild populations. Lions are relatively easy to kill. A small sachet of poison in the hands of a disgruntled villager or organized poacher can quickly kill a whole pride of lions. Evidence suggests that illegal wildlife traders will pay $1,000 for a lion carcass. This kind of payoff far outweighs anything local people currently earn from legally sanctioned use of wildlife, whether trophy hunting or photographic tourism. The danger is that poaching for lion bones could become as catastrophic as illegal trade in elephant ivory or rhinoceros horn.

Hunting is an integral part of Africa's conservation history and its approach to wildlife management. To disentangle hunting from modern African conservation will require a realignment of conservation policy, entrenched since colonial times and embraced and supported by African elites and political interests. But as a civilization that has the ingenuity to put people and machines into space, split the atom, and routinely send unimaginable amounts of information through the ether, surely we can think of a better way to save the wild animals we love besides killing them.

# FOURTEEN

# OUT OF SPACE, OUT OF TIME

The road turns northwest, leaving the hustle and dust of the city of Bulawayo behind. The tarmacadam road cuts across the teak forest of western Matabeleland, tracking the route taken more than a hundred years ago by Allan Wilson and his men as they marched to their doom at the hands of the last warriors of the Ndebele nation. It leads to Hwange National Park and on to the Victoria Falls. Over the last twenty years I've traveled it countless times. Every time I do, I get the feeling of coming home.

In August 1940 my grandparents, Ouma and Oupa Loveridge, bumped along this road in an old eight-horsepower Ford sedan on their way to a honeymoon spent tiger-fishing on the Zambezi at Katambora rapids. They recorded their trip with an 8mm cine camera. The countryside in the flickering footage hardly looks different from my recollections of the landscape from my first visit here half a century later. There was one change: the road, built on a tight budget between the world wars, then consisted of only two strips of tar. In the unlikely event that a driver met with any oncoming traffic, road etiquette dictated that the right-hand strip of tar should be relinquished to the approaching vehicle.

A sign post on the road to
Hwange National Park.

Today, the road is fully tarred, mostly straight and, for this part of Africa, well maintained. But even now, there is little traffic. The three-hour drive to Hwange Main Camp is a time for reflection, a time to relax after stressful meetings with officials in government offices or an endless search for spare parts for our decaying fleet of research vehicles. There is only one settlement before the national park. Lupane is not much more than a village with few administrative offices and a dilapidated fuel station that seldom has fuel, but instead its grubby kiosk offers cans of warm Coca-Cola to thirsty travelers. From Lupane, the road crosses a low, sandy ridge that surveys a valley dotted with traditional huts. Legend has it that Tokoloshi, the mischievous and occasionally malevolent sprite of African mythology, lives here with his wife. I'm not superstitious, but could it be a coincidence that, at this very spot, a haulage truck once sideswiped my research vehicle? Was it pure chance that, on another occasion, my vehicle's front tire blew out, causing me to careen to a stop on the road's gravel shoulder?

The road continues through forests of teak trees, gorgeously green, decked out after the rains in mauve-pink flowers and, in winter, in vibrant yellow and orange leaves. There is a stillness about this wilderness, and I love driving through it for miles and miles without seeing another car or person. Several of the young male lions we have tagged over the last decade have dispersed through these forests, searching for new territories in which to settle. Lobengula, Cecil's eldest son, did so five years ago, moving 150 kilometers to the west of Hwange before doubling back and returning to the sanctuary of the park.

Like an old friend, a lone baobab tree signals the end of the long, hot journey. Perched on a ridge overlooking the Gwaai River, the ancient tree marks the turnoff to Skwatula's old homestead and the dusty road to the Dete Valley, where the lion project started and Stumpy Tail lived with her family a decade and a half before. Musing on the many times I have traveled this way made me realize that roadside Africa had changed a lot in the last twenty years. The forests are still there, but loggers and illegal harvesters of firewood are gradually thinning them. Small subsistence farms have sprung up along the road—patches of cleared land where determined people scratch out meager livings. Thatched, mud, and daub huts now huddle where there was once forest. Ragged kids sell wild fruit and firewood from the roadside. Cattle and goats play chicken with the passing motor traffic. The road is no longer as empty as it once was. It is now an artery through which Africa's rich resources are draining to commercial harbors on the coast, bound for the industrial giants of the East. Thundering trucks, called *Gonyets* by the locals, drag trailers southward, carrying coal and smelted copper ingots on their groaning steel backs. Occasionally they smash through herds of sheep and goats standing in the road and sometimes, at night, lions.

Lions are very much a species caught in the headlights of advancing development. There may now be as few as 20,000 wild lions across the entire African continent, occupying only 8 percent of their historical range.[140] Their numbers are thought to have declined by 43 percent in the last three lion generations—that is, in the last twenty years—about the same time that I have been studying them. There are far fewer lions now than there are African elephants[141] and about the same number as there are white rhinoceroses[142]—both species whose futures worry conservationists. Against this dismal backdrop there are still several secure lion populations, centered in large, well-protected national parks and reserves. Hwange is one of these lion strongholds. The number of lions there has actually increased in the last decade, mostly as a result of improved management and better protection—and, I like to believe, the efforts of our research team.

What is the cause of this drastic collapse in lion numbers? Africa is a continent in flux; its savannah environment is changing faster than it has since the end of the last glacial period 10,000 years ago. Disney has prompted us to imagine Africa as a bucolic scene of endless herds of wildebeests serenely munching on the boundless grasslands, while lions

Cecil's cubs, May 2015.

loll contentedly in the shade of flat-topped acacias. I love *The Lion King* too, but lions don't really hang out with warthogs, eating candy-colored beetle larvae. Sure, hyenas will happily snack on an unprotected lion cub. But the real threat to lions is from people, not hyenas. The pride lands that Mufasa and Simba so proudly surveyed are fast being eroded by an advancing agricultural frontier—its dancing giraffes replaced by cattle and maize fields.

Africa's human population is growing exponentially. Over the next fifty years, it is expected to double from one to two billion people.[143] There are already more than four times as many people here than there were in the 1940s when my grandparents drove through the empty spaces of western Zimbabwe (then Rhodesia). The population has close to doubled since I started working in Hwange in 1994. There seems to be a lot less wildlife too. I would often see kudu and warthogs by the side of the road—occasionally an elephant or giraffe. Now the bush seems emptier. Growing human populations have seen increasing development. Just outside the park, the Gwaai River Hotel used to be something of a frontier outpost. Its barroom, where glassy-eyed, taxidermied trophy heads gazed from the wall, was a place where old-time ranchers, adventurers, hunters, and wildlife officers would come to drink, boast, and fight. Now a township has sprung up nearby. A Chinese coal-mining company built a compound there. Curious visitors are discouraged from nosing about. The plan is apparently to strip-mine part of the Gwaai wildlife area bordering Hwange National Park for its underlying coal deposit. The coal will feed a coal-fired power station to be built right on the national park boundary.

African countries need resources to feed their growing economies. The continent's natural wealth is in demand too. There is a market abroad for everything from minerals to rhinoceros horn, timber to pangolin scales. The growing human population needs space—space to live, space to grow food, space for domestic livestock. Hungry people view wild animals, not with wonder but as free protein to be harvested. By 2060 to feed the human population of Africa, it is estimated that 430 million hectares of land, an area half the size of the continental United States, will need to be cleared for agriculture.[144] Much of this land will be cleared within parts of Africa that are currently inhabited by lions. Under this scenario, lions will only continue to exist in the largest protected areas. So unless fragile habitat links between protected areas are conserved, like a spiderweb of safety lines preventing their total isolation, then the descendants of Dynamite will no longer be able to journey across the landscape, colonize new areas, and carry their genes into adjacent populations.

Given the burgeoning human populations in Africa, loss of many formerly pristine habitats outside protected areas is likely but is not inevitable. Agriculture in Africa is often inefficient subsistence farming, requiring far more land than modern intensive cultivation. Use of more efficient farming methods and appropriate crops and technologies could actually reduce the future demand for agricultural land by 55 percent, allowing potential wildlife land to be spared. The future of lions and other wildlife may largely depend on whether this can be achieved.

Lions are one of the most beloved animals on the planet. Da Vinci, Rembrandt, and Landseer harnessed the big cat's image in art to portray power, majesty, and strength. They are the national symbol of no fewer than fifteen countries, many of which do not themselves have extant populations of lions. This is nowhere so obvious as in the United Kingdom, where heraldic images of lions adorn the paraphernalia of state and military institutions and national sports teams. Yet it is close to 12,000 years since cave lions stalked Pleistocene Britain. The contrarian might muse on whether humanity needs to protect predators such as lions in the wild since it is readily apparent that their cultural value persists even once the beast itself has gone. Since large predators can be troublesome creatures for people to live with, some might say it is enough that they are represented in zoos. After all, lions and tigers are likely to be public attractions for as long as zoos exist. Who cares whether they survive in places few people will ever visit?

Unsurprisingly, I'd argue that loss of any species in the wild, let alone spectacular ones such as lions, is an act of ecological vandalism closely akin to the deliberate destruction of priceless works of art or the demolition of cultural artifacts by religious extremists of the Islamic State and Taliban. However, there may be more utilitarian reasons to keep predators around. The presence of charismatic predators in some ecosystems can be economically valuable, particularly where economies are dependent on tourism earnings.[145] But there are even more fundamental reasons for sparing them. In the middle of the last century, ecologists started to realize that elimination of the top predators from ecosystems destabilized intricate webs of species relationships, with far-reaching consequences.[146] There is growing evidence that predators regulate the populations and eating habits of their herbivore prey. They indirectly influence vegetation structure, and in turn the flow of rivers

and the frequency of fire—thus storing carbon and buffering climate change. They control the spread of zoonotic diseases and help ecosystems resist catastrophic change.[147] Eliminating the apex predators from an ecosystem is like ripping out the carefully shaped keystone from an archway—causing it to collapse in on itself.

The death of Cecil the lion caused the largest global reaction to an animal story ever. It's clear that one lion can motivate people to care about the fate of its species. It suggests that the lion could serve as a valuable ambassador for African ecosystems. Lions need enormous spaces and substantial wild prey to survive in anything approximating a natural state, so to conserve lions you have to conserve it all.

What will it take for this to happen? Like much in life, it all comes down to cost. The first reality is that most African states cannot afford the luxury of conserving wild animals, nor do their people necessarily want to live alongside them or accept the opportunity costs of setting aside land for conservation. It seems unfair to burden the people of

Africa, some of the poorest in the world, with the costs of protecting species that are globally important and internationally valued. Somehow conservation needs to be globally subsidized. There are already models in place. Wealthy developed countries aid developing countries. Most of this aid pays for much-needed development, humanitarian assistance, and poverty alleviation, though much less money goes toward environmental management and protection. The budget deficit for management of protected areas within the African lion range is estimated to be $380 million a year.[148] To put this in perspective, this is about what it costs to purchase a single F-22 Raptor fighter jet. This seems a sad prioritization and perhaps one that future generations will regret if it turns out their forebears chose destruction over the protection of the priceless natural environment.

Increasingly, private philanthropy funds conservation. One highly successful example of this is the nonprofit conservation organization African Parks. This organization, founded in 2000, partners with governments and local communities to rehabilitate and manage African protected areas.[149] So far this nonprofit has taken over the management of twelve parks across Africa and aims to increase the number of protected areas under management to twenty by 2020. In 2016 African Parks spent $34.8 million on conservation management, much of this funded directly by donations. These initiatives, which are effectively privatization of conservation to reduce the burden on poor African states, hold significant promise.

In a piece published in the journal *Animals*,[150] several colleagues suggested how funding conservation of iconic species might be implemented. They observe that images of species such as lions are used as logos on many commercial and corporate brands. If small donations were made for each sale, significant funds could quickly amass. For example, they suggest that if £1 were added to the sale of each football shirt sold by the English Premier League (which uses a crowned lion as its mascot), enough funding could be raised per year to employ 4,000 lion guardians in Africa. In another example, they note that in the British egg industry, most eggs are stamped with a quality-assurance stamp in the form of a British lion. Because around 34 million eggs are consumed in Britain each day, a modest premium charged on each stamped egg would generate more than £10 million a year and fund a national lion conservation program in Africa. Maybe this sounds like pie in the sky. But perhaps this kind of self-imposed tax on the use of the image of a culturally relevant animal is not so far-fetched. If several years ago you had asked me whether the death of one of the Hwange study lions could have generated global concern and headlines across the world, I'd have laughed out loud.

In October 2017 I watched as a pride of lions squabbled over the carcass of an elephant they had recently killed. One of the young males, a three-year-old with an attitude, was particularly vocal as he defended the prime piece of rib cage he was gnawing on. His scrubby mane bristled with adolescent indignation as he grumbled and swatted at his brother and five sisters. Nobhule and Sissy, the two big lionesses of the Ngweshla pride, looked on tolerantly as their offspring bickered. It was thanks to their experience and fierce protection that the seven youngsters had survived at all. For these are the almost-grown cubs of Cecil, the last litter he sired before he was killed. Back then, in July 2015, with their father dead, we had little hope they would survive the infanticide that almost always follows the death of a pride male. Against all odds, they had grown and prospered. As I watched them, I wondered about their future. Cecil's two youngest sons are just about at an age when they will disperse and run the gauntlet of dangers all young lions must face. They do so in an Africa that is changing forever. Their survival and that of their descendants, if they have any, may depend on whether we can reevaluate our relationship with nature such that wilderness, lions, and other wild animals can be recognized as the priceless treasures they are and their protection regarded as a global responsibility.

# ACKNOWLEDGMENTS

The germ of this book was conceived in July 2015 at the height of the Cecil affair, when amongst the thousands of emails I received from the media and general public, there was one from book editor Stacy Creamer, who suggested that I might write a book about Cecil and lion conservation. Her continued enthusiasm for this project made this book a reality. The support and guidance of Judith Regan, Kathy Huck, Lynne Ciccaglione, and George Rush brought this book to fruition.

I am fortunate to hold a research fellowship at Lady Margaret Hall and the Wildlife Conservation Research Unit, Department of Zoology, University of Oxford. In Oxford, friends and colleagues in the Wild-CRU, LMH, and the wider university community have shaped my thinking, provided a warm camaraderie, and made Oxford my academic home. In particular, my friend and mentor, Professor David Macdonald, has guided and supported my career as a conservation biologist.

None of the research and conservation work on lions in Hwange would have been possible without funders and donors who support the WildCRU's Hwange Lion Project: the Darwin Initiative for Biodiversity, Mitsubishi Corporation Fund for Europe and Africa, R. G. Frankenberg, Boesak and Kruger, Rufford Maurice Laing, Disney Conservation Fund, Eppley, Panthera, Robertson and Recanati-Kaplan Foundations, SATIB Trust, Riv and Joan Winant, and particularly the generosity of

Tom and Daphne Kaplan, and the members of the public whose donations formed the Cecil Fund, which supports ongoing lion conservation work and the training of Zimbabwe's next generation of conservation professionals.

In Hwange, I thank my many friends and colleagues in National Parks who have facilitated, assisted with, and shared in the lion research work. Andrew Searle, Lionel Reynolds, and Alan Elliott made the lion research project possible. Jane Hunt has been the backbone of the project for a decade and a half. Its past and continued success would not have been possible without project research staff and doctoral students: Lindy Denlinger, Bill Eldridge, Donna Robertson, Zeke Davidson, Kate Smith, Dan Parker, Pete Roberts, Joanne Loveridge, Agrippa Moyo, Paul Bennett, Senga Brady, Dennis Lander, Roger Lewis, Chris Freeman, Nic Elliot, Jaelle Claypole, Brent Stapelkamp, Laurie Simpson, Matt Wijers, Byron duPreez, Jean Purdon, Harley Peacock, Moreangels Mbizah, Paul Trethowan, Nicholas Long, Lowane Mpofu, Liomba Mathe, Andrea Sibanda, Lovemore Sibanda, and Justin Seymour-Smith. I am indebted to Andrea Sibanda for the use of his notes taken during the investigation of the Cecil hunt and to Jane and Andrea for collating and writing down the stories surrounding the death of Elton Ndlovu.

# NOTES

## PROLOGUE

1   The name Cecil was readily adopted by the local safari guides and their tourist visitors. However, once "Cecil the lion" became a posthumous front-page celebrity, some erroneously assumed the name commemorated the arch-colonialist Cecil John Rhodes. Rhodes led the invasion by a paramilitary force and colonization of the land that was to be named Rhodesia, and renamed Zimbabwe after independence from Britain in 1980. People in southern Africa are justifiably sensitive about the region's colonial past. Of course it was never our intention to make a political statement.

2   "Cecil" shared the podium with "Tilikum," the Orca whale, and "Uncle Sam," the American bald eagle who famously attacked Donald Trump during a photo shoot. Stein J. "Four legs good, two legs irrelevant. *Time* magazine's 100 most influential animals." *Time*. May 2, 2016.

## CHAPTER 1

3   Zimbabwe's fauna, flora, and protected areas are managed and protected by the Zimbabwe National Parks and Wildlife Management Authority (formerly the Department of National Parks and Wildlife Management). In this book I will refer to the organisation as National Parks, Zimbabwe National Parks, or the National Parks Authority.

4   Sound designer Mike Mangini admitted to using the sounds recorded from a tiger to produce the "'roar' given by MGM mascot 'Leo.'" www.woot.com/blog/post/the-debunker-what-kind-of-animal-roars-at-the-start-of-the-mgm-movies.

5   A "pan" is the colloquial southern African name for a seasonal water hole. There are thousands of seasonal pans across the Kalahari system, holding water for several months, sometimes for the entire year. Some pans in Hwange National Park are artificially supplemented with water from boreholes to provide a permanent water source for wildlife.

6   Biltong is dried, salted, and spiced meat. It is the southern African equivalent of beef jerky, though biltong is often made with game meat. It was originally made by hunters and farmers as a means of preserving meat in a hot climate before the advent of refrigeration.

7   A firsthand account of this episode is given by Willem De Beer in *National Parks and Wildlife Management: Rhodesia and Zimbabwe 1928–1990*. Editor M. Bromwich. ISBN 978-0-620-61929-5. My father spent time in Wankie in 1972 and 1973, catching crocodiles with National Parks colleagues. He knew Willem De Beer and likely heard the story of the Shapi lion firsthand from him and other colleagues working in Wankie at the time.

8   Galamine triethiodide (Flaxedil) is still the drug of choice for immobilizing large crocodiles. My father, John Loveridge, and his colleague Dave Blake experimented with this drug and calculated the dosage rates for capturing crocodiles of different sizes. "Techniques in the immobilisation and handling of the Nile crocodile, *Crocodylus niloticus*." Loveridge, J.P, Blake, D.K. 1972. *Arnoldia*, Volume 5 (40) (1972), p. 14.

9   Operation Noah took place between 1958 and 1964 as Kariba hydroelectric dam on the Zambezi River flooded the middle Zambezi Valley. Rupert Fothergill led the rescue of more than 6,000 wild animals marooned on newly formed islands. The motor launch *Erica* was a private donation to assist with the operation.

10   Records suggest that tens of thousands of wild crocodiles were hunted for their skins in the late 1950s and 1960s. Up to 20,000 crocodiles were hunted in the Oka-vango Delta between 1957 and 1969. Four thousand crocodiles were hunted in the mid-Zambezi in a two-year period in the late 1950s before the Kariba Dam was completed. In: "Crocodile numbers in Lake Kariba, Zimbabwe and factors influencing them." Taylor, R., Blake, D. and Loveridge, J.P. 1992. Proceedings of the 11th working group meeting. IUCN Crocodile Specialist Group.

11   Hwange National Park is 14,500 square kilometers. The area was designated as the Wankie Game Sanctuary in February 1928. The original ordinance listed antelope, elephant, rhino, hippo, and ostrich as being protected. Predators were not protected. The first warden of the park was Ted Davison. The area was proclaimed a national park in 1950.

12   A *stoep* is a deep veranda or porch found on many older farmhouses in Africa, providing a shaded outside space for times of the year when the sweltering heat makes living indoors impossible.

## CHAPTER 2

13   "A method of identifying individual lions (*Panthera leo*), with an analysis of re-liability of identification." Pennycuick, C. J. and Rudnai, J. 1970. *Journal of Zoology* Volume 160, pages 497–508.

14   "Reproductive success in lions." Packer, C., Herbst, L. Pusey, A. Bygott, D., Hanby, J. Cairns, S., Borgerhoff-Mulder, M. In: *Reproductive Success.* Clutton-Brock, T. (ed.). 1988. University of Chicago Press, Chicago.

15   In the Serengeti in more than 60 percent of coalitions consisting of two males, the males are unrelated; more than 50 percent of trios contain an unrelated male. However, males in groups of four or more were invariably closely related. "A molecular genetic analysis of kinship and co-operation in African lions." Packer, C., Gilbert, D., Pusey, A., O'Brien, S. Nature, 1991, Volume 351, page 562.

16   "Roaring and numerical assessment in contests between groups of female lions." McComb, K., Packer, C., Pusey, A. 1994. Animal Behaviour. Volume 47, pages 379–387.

17   Antoinette is one of the small holdings set along the length of the Bulawayo to Victoria Falls railway line when it was built in the 1920s. The original intention had been to encourage settlers to live along the railway. One of their obligations was to pro-vide a ready supply of water for the steam trains that plied the line. However, the farms were too small to be economically viable. Few were permanently settled. Measuring five by five kilometers, Antoinette is tiny in relation to the range of a pride of lions, and yet, over the years, more lions have been killed by trophy hunters there than anywhere else in the Hwange area. In 1999 and 2000 the owners of the concession requested approval from wildlife authorities for a lion-hunting quota of two males and two females, which they duly received. Over the eighteen years, we have recorded a tally of twenty lions (30 percent of lions trophy-hunted during the study) having been killed by trophy hunters on that one tiny bit of land. The most famous was a lion named Cecil.

## CHAPTER 3

18  "Job-Related Mortality of Wildlife Workers in the United States," 1937–2000. Blake-Sasse, D. *Wildlife Society Bulletin*. 2003. Vol. 31, No. 4, pp. 1015–1020.

## CHAPTER 4

19  "People and Wild Felids: Conservation of cats and management of conflicts." Loveridge, A.J.; Wang, S., Frank, L., Seidensticker, J. 2010. In: *Biology and Conservation of Wild Felids*. Macdonald, D.W. and Loveridge, A.J. (eds.). Oxford University Press.

20  Between 2006 and 2010, we recorded the length of galvanized fencing wire that was stolen from the 72-kilometer Foot and Mouth veterinary fence on the eastern border of Hwange National Park. Over five years 285 kilometers of wire was removed. If all of this wire was used to produce snares (on average 3 meters of wire used in each), then more than 71,000 snares could have been manufactured. As a consequence, this part of the national park was a hot spot of poaching over this period. "The landscape of anthropogenic mortality: how African lions respond to spatial variation in risk." Loveridge, A.J., Valiex, M., Elliot, N. & Macdonald, D.W. 2017. *Journal of Applied Ecology* Volume 54, issue 3, pp815–825.

21  Spotted hyena noses are particularly prized by practitioners of traditional medicine, as they can be used to "sniff out" witchcraft and curses cast by enemies. Lion and leopard claws, teeth, and skin are sold to local consumers as well as to unsuspecting tourists looking for curios. Growing demand for big-cat body parts and bones for use in in traditional Asian medicines has exacerbated the poaching of big cats for sale to illegal wildlife traders.

22  The project has rescued many lions from snares. In every case, despite the brutal nature of the injury, the animals have all recovered remarkably well once the wire has been removed from the wound.

23  In Hwange, over eighteen years, we have recorded 234 mortalities. Of those deaths, 77 percent were caused directly by people. Very few of our study animals have died of old age. Trophy hunting accounts for 30 percent of mortality (48 percent of male mortalities recorded); deaths in poacher's snares: 19 percent; retaliatory killing when in conflict with people: 23 percent; hit by trains or motor vehicles: 5 percent. For further details see: "The impact of sport hunting on the population dynamics of an African lion population in a protected area." 2007. A. J. Loveridge, A. W. Searle, F. Murindagomo, D. W. Macdonald, *Biological Conservation* 134, 548, and "Conservation of large predator populations: demographic and spatial responses of African lions to the intensity of trophy hunting." Loveridge, A.J., Valiex, M., Davidson, Z., Mtare, G. & Macdonald, D.W. 2016 *Biological Conservation*. Volume 207. pp 247–254.

## CHAPTER 5

24  Hunting quotas (the maximum number of animals that it is estimated can be sustainably harvested from a population) are allocated by the wildlife authority, usually National Parks, based on wildlife population estimates provided by either landowner or park ecologists. In the case of landowners, wildlife population estimates are often either overestimated or inflated. As the allocated hunting quota is used as the basis for marketing trophy hunts, there is clearly a financial incentive to obtain as high a quota as possible.

25    The lion-hunting quota given in the Gwaai Conservancy in these years (1999–2003) was 2.5 lions per 100 square kilometers. This is fifty times higher than the recommended sustainable offtake from a lion population of 0.05/100km². Actual hunting offtakes were 0.18 lions per 100km², close to four times more than is considered sustainable. "The impact of sport hunting on the population dynamics of an African lion population in a protected area." 2007. A.J. Loveridge, A. W. Searle, F. Murindagomo, D. W. Macdonald. *Biological Conservation* 134, 548.

26    "Conservation of large predator populations: demographic and spatial responses of African lions to the intensity of trophy hunting." Loveridge, A.J., Valiex, M., Davidson, Z., Mtare, G. & Macdonald, D.W. 2016 *Biological Conservation*. Volume 207. pp 247–254.

27    Lion hunting was suspended from 2005 to 2008 in northwestern Matabeleland, including the hunting concessions around Hwange. Despite pressure from hunters, Zimbabwe National Parks maintained this hunting moratorium, reintroducing hunting at much lower levels in 2009. In response to lobbying from the hunting industry, a spokesperson for the National Parks and Wildlife Management Authority, retired Major Mbewe, stated that the ban was still in force and that he didn't "think the numbers are good enough to start hunting. We still have a few lions and their numbers need to grow first." Article: "Zimbabwe: Lion hunt ban still on." 8/16/2008. Radio Voice of the People.

28    *Lions in the Balance: Man-eaters, Manes and Men with Guns.* Packer, C. 2015. University of Chicago Press.

29    Details of the recovery of the Hwange lion population after trophy hunting was suspended can be found in: "African Lions on the Edge." A. J. Loveridge, G. Hemson, Z. Davidson, D. W. Macdonald. 2010. In *Biology and Conservation of Wild Felids*, D. W. Macdonald, A. J. Loveridge (Eds.). Oxford University Press, Oxford, pp. 283–304.

30    The nefarious activities of hunting company Out of Africa Adventurous Safaris and its owners are widely documented in the media. For example:

"US hunter sentenced for illegal leopard trophy." Pierre, S. October 17, 2008. www.thedenverchannel.com/news/17749641/detail.html.

"Zanu-PF poaching links exposed." Ndlovu, R. and Groenewald, Y. October 08 2010. *Mail and Guardian*, accessed 10/13/10, https://mg.co.za/print/2010-10-08-zanupf-poaching-links-exposed.

"Rhinos: Controversial safari operator held." Julian Rademeyer and Marietie Louw-Carstens, September 21, 2010. *Beeld*. www.news24.com/SouthAfrica/News/Rhinos-Controversial-safari-operator-held-20100921.

31    The plunder of wildlife resources in the Gwaai Conservancy and elsewhere in Zimbabwe has been extensively documented by Zimbabwean investigative journalist Oscar Nkala on his blog African Environmental Police. Including posts:

"Dawie Groenewald, the man that so loved Zimbabwe that he plundered all its rhinos and elephants," May 13, 2011. (africanenvironmentalpolice.blogspot.co.uk/2011/05/dawie-groenewald-man-that-so-loved.html).

"Zimbabwe Poaching Report Part Two." May 15, 2011 (africanenvironmentalpolice.blogspot.co.uk/2011/05/Zimbabwe-paoching-report-part-2.html).

32   "Impacts of trophy hunting on lions in East and Southern Africa: Recent offtake and future recommendations," The Eastern and Southern African Lion conservation Workshop, Johannesburg, South Africa. Packer, C., Whitman, K., Loveridge, A., Jackson, J., Funston, P., 2006. Background paper presented at IUCN Lion Conservation Workshop, Johannesburg. 11–13 January 2006.

## CHAPTER 6

33   Statistically, lions and hyenas attacked livestock wearing bells more frequently than would be expected, based on the proportion of animals fitted with bells. "Bells, bomas and beefsteak. Complex patterns of human predator conflict at the protected area-agropastoral interface." Loveridge, A. J., Kuiper, T., Parry, R., Sibanda, L., Stapelkamp, B., Hunt, J. E., Sebele, L., Macdonald, D. W. (2017). *PeerJ, 5:e2898.*

34   For the purposes of telling this story I have amalgamated several genuine conversations with villagers about the issue of lion depredation. Mr. Nkomo is a fictional character but is based on a real person, an elder from Ziga village in Tsholotsho. The real-life old gentleman held similarly strong opinions and expressed the same fluent skepticism of the research we were doing. Jane and I had several conversations with him of the kind portrayed. I have chosen to use the surname Nkomo, as it is a common Ndebele name, which translates to "cow," so Mr. Nkomo is literally Mr. Cow, which seemed appropriate.

35   Only forty-three of the attacks, 6 percent, were on livestock secured in bomas. The vast majority of attacks (559) occurred on livestock left unprotected in grazing areas at night. "Bells, bomas and beefsteak. Complex patterns of human predator conflict at the protected area- agropastoral interface." Loveridge, A. J., Kuiper, T., Parry, R., Sibanda, L., Stapelkamp, B., Hunt, J. E., Sebele, L., Macdonald, D. W. 2017. *PeerJ, 5:e2898.*

36   "Seasonal herding practices influence predation on domestic stock by African lions along a protected area boundary." Kuiper, T., Loveridge, A.J., Parker, D., Hunt, J.E., Stapelkamp, B., Sibanda, L. & Macdonald, D.W. 2015. *Biological Conservation,* 191, 546–554.

37   Of twenty-three lionesses killed for eating livestock, 82 percent were adults; 22 percent of males killed in conflict were adults. The vast majority of male lions killed for eating livestock are dispersal-age subadults.

38   In 353 households surveyed, 51 percent of respondents ranked fencing the park as the most favored solution to conflict with wild animals; 81 percent ranked fencing in their top three most favored solutions.

39   "Conserving Large Carnivores: Dollars and Fence." Packer, C., Loveridge, A.J., Canney, S. et al. 2013. *Ecology Letters* 16, 635–641.

## CHAPTER 7

40   Gary Haynes, 2010. "Puzzling over the Bumbuzi spoor engravings in Zimbabwe." Unpublished. Accessed at www.researchgate.net/publication/267545631_Puzzling_Over_The_Bumbusi_Spoor_Engravings_In_Zimbabwe. 30/08/2017.

41   A. Wannenburg. 1980. *The Bushmen.* Country Life Books. Hamlyn. London. Plate 40.

42  "Lions cause stir in Hwange." Southern Eye. January, 23, 2015. Muponde, R. www.southerneye.co.zw/2015/01/23/lions-cause-stir-hwange.

43  "Lion Conservation in Tanzania Leads to Serious Human–Lion Conflicts." R. Baldus. 2004. Tanzania Wildlife Discussion Paper No. 41. Unpublished.

44  In Tanzania, 563 people were killed and 307 injured by lions between 1990 and 2004. "Lion attacks on humans in Tanzania." C. Packer et al. 2005. Nature 436: 927–928.

45  In his book about man-eating lions in Kruger National Park, Robert Frump estimates that 13,380 Mozambican refugees may have been killed and eaten by lions between 1960 and 2005. While there is no evidence to verify this figure, it is likely that many refugees were killed by lions with only a fraction of cases actually recorded. Maneaters of Eden. Life and Death in Kruger National Park. R. Frump. 2006. The Lyons Press. Connecticut.

46  D. Grobler. 2003. "Lions change by darkness." African Lion News 4. www.african-lion.org.

47  The story appeared in the Bulawayo Chronicle under the "Weird News" section on page three with a strapline of "Boy (7) killed by lion" The story appeared alongside the fatuously embellished headlines: "Prostitute strangles client" and "I'm no prophet: Magistrate." Sunday News Reporter. Sunday Mail, Bulawayo Chronicle. April 6, 2014, p 3.

48  Indeed, one Zimbabwean, Goodwell Nzou, writing for the New York Times, went further. His blog described how many Africans were bemused by the outpourings of grief expressed, in print and social media across the world, at the death of a lion. Most of it by people who had never seen a lion outside a wildlife documentary and had certainly never known this particular lion. Nzou went on to relate how, in the rural village in northern Zimbabwe in which he grew up, not far from where Masweresei once terrorized the community, the death of a lion was cause for celebration and not mourning. The removal of a dangerous man-eater and pestilential killer of livestock was a public service to be applauded. "Zimbabwe, we don't cry for the lions." Nzou, G. New York Times. Opinion pages. http://well.blogs.nytimes.com/2015/08/05.

## CHAPTER 8

49  "Case Study of a Population Bottleneck: Lions of the Ngorongoro Crater." Packer, C. Pusey, A. Rowley, H., Gilbert, D., Martenson, J. and O'Brien, S. 1991. Conservation Biology, Volume 5: Page 219.

50  Lion (Panthera leo) populations are declining rapidly across Africa, except in intensively managed areas. Hans Bauer, Guillaume Chapron, Kristin Nowell, Philipp Henschel, Paul Funston, Luke T. B. Hunter, David W. Macdonald, and Craig Packer. Proceedings of the National Academy of Sciences, 2015, Volume 112, pg. 14895.

51  A Hunter's Wanderings in Africa. Frederick Courteney Selous. 1881. Rhodesiana Reprint Library Volume 14, Facsimile of the 1881 edition. Page 257.

52  Correspondence in the Sunday Times (South Africa), 1908, criticizing management of Sabi Game Reserve. Quoted in G. L. Smuts. Lion, pg 176.

53  Lion. G.L Smuts, 1982. Macmillan, Johannesburg, South Africa. Page 187.

54  Wankie. The Story of a Great Game Reserve. E. Davison. 1967. Thorntree Press reprinted 1998.

55   "The Lion in West Africa Is Critically Endangered." Philipp Henschel, Lauren Coad, Cole Burton, Beatrice Chataigner, Andrew Dunn, David Macdonald, Yohanna Saidu, Luke T. B. Hunter. *Public Library of Science, One.* Volume 9, issue 1, e83500.

56   In collaboration with landscape ecologist Sam Cushman, from the US Forest Service, we have developed spatial models to predict how lions might be able to move across the southern African KAZA landscape, to provide governments and wildlife conservation policy makers with information for planning conservation interventions in parts of the landscape that are critical for the long-term survival of lions. Our modeled predictions show that if we only protect existing national parks, the connectivity of the landscape—in terms of a lion's ability to move across it—will be reduced by more than 50 percent, dangerously fragmenting lion populations. It is essential that other wildlife areas, such as state forests and hunting areas, are also protected, as these form the critical corridors between the fully protected national parks. There may also be a need to establish corridors through some farm areas, as yet sparsely populated by people, in order to offer lions bridges from smaller protected areas to the immense national parks and game reserves in western Zimbabwe and Botswana. Our models show the most critical corridors are those that connect Hwange with Chizarira, the corridor we know that Hwange lions have used in the past, and also the linkages between the lion populations of Hwange and northeastern Botswana and the vast Central Kalahari. "The devil is in the dispersers. Predictions of landscape connectivity change with demography." Elliot, N., Cushman, S.A., Macdonald, D.W. & Loveridge, A.J. 2014. *Journal of Applied Ecology*, 51, 1169–1178. A multiscale assessment of population connectivity in African lions (*Panthera leo*) in response to landscape change. Cushman, S.A., Elliot, N., Macdonald, D.W. & Loveridge, A.J. 2015. *Landscape Ecology*, DOI 10.1007/s10980-10015-10292-10983.

## CHAPTER 9

57   George Schaller. 1972. *The Serengeti Lion*. A study of predator prey relationships. Pg 121.

58   They are also quite capable of moving large distances over short periods. I once recorded a male lion covering sixty kilometers in a single night.

59   Paul Trethowan, a WildCRU doctoral student, fitted custom-built loggers to lions that measured in minute detail every movement the lions made. By calibrating these against observed activities, we could estimate how much time lions spent hunting and how often they were successful in making kills. "The hunger games. Sex based responses to hunger and hunting in African lions" (*Panthera leo*). Trethowan, P., Hart, T., Loveridge, A. J., Markham, A., Wijers, M., DuPreez, B., Macdonald, D. W. 2016. *Current Biology*.

60   "Serengeti Ecosystem: Profitability, encounter rates, and prey choice of African lions." Scheel, D., 1993. *Behavioural Ecology*, Volume 4: p. 90–97. "Etosha: Hunting success of lions in a semi-arid enviroment." Stander, P.E. and S.E. Albon. In: *Mammals as Predators*. N. Dunstone and M.L. Gorman (Editors). 1993, Symposium of the Zoological Society of London, No 65: London. p. 127–143.

61   Valeix, M., et al., "Behavioural adjustments of African herbivores to predation risk by lions: Spatiotemporal variations influence habitat use." *Ecology*, 2009. 90(1): p. 23–30.Valeix, M., A.J. Loveridge, and D.W. Macdonald, "Influence of prey dispersion and group size of African lions: a test of the resource dispersion hypothesis." *Ecology*, 2012. 93(11): p. 2490–2496.Valeix, M., et al., "Does the risk of encountering lions influence African herbivore behaviour at waterholes?" *Behavioural Ecology and Sociobiology*, 2009. 63: p. 1483-1494.Valeix, M., et al., "Understanding patch departure rules for large carnivores: Lion movements support a patch disturbance hypothesis." *The American Naturalist*, 2011. 178(2): DOI: 10.1086/660824.

62   Elephant calves are killed by lions in the dry months of years with below-average rainfall, suggesting that elephants are more vulnerable at these times, most likely because of shortage of food and limited access to water. Of the 230 elephants we have recorded as lion victims, the majority were pre-adolescent, on average around four years old, but as old as fourteen and as young as one month. "Influence of drought on predation of elephant (*Loxodonta africana*) calves by lions (*Panthera leo*) in an African wooded savannah." Loveridge, A.J., Hunt, J.E., Murindagomo, F., Macdonald, D.W. 2006. *Journal of Zoology* 270.

63   Based on weighed stomach contents. A male lion had eaten thirty kilograms, a ten-month-old cub, 25 percent of its body weight. Humans eat around 4 percent of body weight per day. *Lion*. Smuts, G.L. Macmillan, Johannesburg, South Africa. Page 245.

64   In the algebra of ecology, the herbivore population is determined by the amount of grass or browse available for them to feed on. In African savannahs, this in turn is determined by rainfall and soil richness, but also by the protection from development and exploitation afforded to wildlife. These factors vary across Africa. In the nutrient-rich, wet highlands of East Africa, places like the Maasai Mara or Serengeti, prey is abundant (exceeding 10,000 kilos per square kilometer). Here, protected areas support twenty to thirty lions for every hundred square kilometers of protected area. It requires 350 square kilometers to protect a population of 100 lions. But over much of Africa, soil is relatively poor and rainfall is low and erratic. These areas support far fewer wild prey animals and therefore far fewer lions. At the extremes, in places like the southern Kalahari, prey biomass can be as low as 200 kilos per square kilometer and can support fewer than one lion for every 100 square kilometers. Here lions have vast home ranges simply to ensure that they have access to enough food.

## CHAPTER 10

65   This was an astonishing behavioral observation, and I have only heard of one other case where this has happened. This was in the Serengeti, where a subadult male joined his father's coalition of two (see "Co-operation and competition within coalitions of male lions. Kin selection or Game theory?" Packer. C. and Pusey, A. 1982. *Nature* Volume 296. Pg. 740–742).

66   In a reproductive career spanning eight years, Mpofu sired at least fifty-two cubs in six different prides. Twenty of his offspring survived to adulthood. He sired his last litter of cubs in the Kennedy Pride in July 2008 when he was ten and a half years old. Over his lifetime, Cecil sired at least thirty-four cubs, fifteen of which survived to adulthood. Cecil sired his last litter of cubs in the Ngweshla pride in December 2014 when he was eleven and a half years old. Nine of Cecil's last litter of cubs survived. I last saw them in October 2017.

67   Of the ninety-six adult male lions we have collared and studied over nearly twenty years, close to half have been trophy-hunted. Twenty have been killed as problem animals or by poachers. Only six have died of natural causes. Of these, only one, Jericho, has died of old age.

## CHAPTER 11

68   Based on the Whitman and Packer model that shows if hunted males are six years old or older, the population is stable. "Sustainable trophy hunting of African lions." Whitman, K., Starfield, A. M., Quadling, H. S., & Packer, C. 2004. *Nature*,

428, 175–178. It seems likely that the age threshold is actually significantly higher and that six years is too young. "Assessing the sustainability of lion trophy hunting with recomendations for policy." Creel, S., M'Soka, J., Droge, E., Rosenblatt, E., Becker, M. S., Matandiko, W., & Simpamba, T. 2016. *Ecological Applications*, 26(7), 2347–2357. doi:doi:10.1002/eap.1377.

69   *The Telegraph*, July 31, 2015. "Cecil the lion's killer Walter Palmer 'wanted to stalk an elephant next—but couldn't find one big enough.'" P. Thornycroft and H. Alexander. www.telegraph.co.uk/news/worldnews/africaandindianocean/zimbabwe/11773653/Cecil-the-lions-killer-Walter-Palmer-wanted-to-stalk-an-elephant-next-but-couldnt-find-one-big-enough.html.

70   Quotas are set to ensure sustainability of a local population; shooting an animal from one population and claiming to have been hunted from a different population defeats the purpose. Worryingly, it allows animals to be illegally hunted in places where no hunting quotas have been issued and the trophy to be subsequently "laundered" through a hunting area where quotas have been approved. As well as being dishonest, this manipulation of area-specific quotas potentially endangers populations. This is exactly what appears to have happened in the case of the killing of Cecil.

71   *The Telegraph*. July 28, 2015. Alexander, H. Thornycroft, P. Laing, A. "Cecil the lion's killer revealed as American dentist." www.telegraph.co.uk/news/worldnews/africaandindianocean/zimbabwe/11767119/Cecil-the-lions-killer-revealed-as-American-dentist.html.

72   *The Telegraph*. July 28 2015. Alexander, H. Thornycroft, P. Laing, A. "Cecil the lion's killer revealed as American dentist." www.telegraph.co.uk/news/worldnews/africaandindianocean/zimbabwe/11767119/Cecil-the-lions-killer-revealed-as-American-dentist.html

73   *ABC News*. August 13, 2015. Sheheri, T., Berman, T. Valiente, A. "See Photos of Black Bear Illegally Hunted by Dentist Walter Palmer Who Killed Cecil the Lion." http://abcnews.go.com/US/photos-black-bear-illegally-hunted-dentist-walter-palmer/story?id=33067963.

74   According to US media reports, Palmer's permit stipulated the hunt should take place in Washburn County, Zone A1. Instead, he hunted the bear in Price County, Bear Management Zone A. *Fox 9 News*. Lyden, T. September 14, 2015. "Guide: Walter Palmer knew Wis. black bear 'was illegally shot and killed.'" www.fox9.com/news/guide-walter-palmer-knew-wis-black-bear-was-illegally-shot-and-killed.

75   *The Guardian*. August 14, 2015. "Hunter who shot Cecil the lion illegally killed black bear in Wisconsin in 2006." www.theguardian.com/us-news/2015/aug/14/walter-palmer-illegally-killed-bear-wisconsin-cecil-the-lion.

76   *ABC News*. August 13, 2015. Sheheri, T., Berman, T. Valiente, A. "See Photos of Black Bear Illegally Hunted by Dentist Walter Palmer Who Killed Cecil the Lion." http://abcnews.go.com/US/photos-black-bear-illegally-hunted-dentist-walter-palmer/story?id=33067963.

77   *The Telegraph*, July 31, 2015. "Cecil the lion's killer Walter Palmer 'wanted to stalk an elephant next—but couldn't find one big enough.'" P. Thornycroft and H. Alexander. www.telegraph.co.uk/news/worldnews/africaandindianocean/zimbabwe/11773653/Cecil-the-lions-killer-Walter-Palmer-wanted-to-stalk-an-elephant-next-but-couldnt-find-one-big-enough.html.

78   For example, Pope and Young Club (a US hunting association that promotes bow hunting) record book entry requirements stipulate that the hunted animal must have been hunted "entirely by the use of the bow and arrow"; https://pope-young.org/records/entry_requirements.asp. Other hunting clubs have similar requirements for records of "bow hunted" trophies.

## CHAPTER 12

79   The photograph was in fact of Cecil and was taken by photographer Vincent O'Sullivan, www.flickr.com/photos/vjosullivan/14687370670.

80   Press statement, Beks Ndlovu, CEO. African Bush Camps. www.africanbushcamps.com. July 15, 2015.

81   Accurate Reloading: African Hunting Forum. http://forums.accuratereloading.com/eve/forums/a/tpc/f/1411043/m/9551038212.

82   A good example of some of this dialogue can be found on the web forum Safaritalk. http://safaritalk.net/topic/14828-hwange-national-park-press-release-beks-ndlovu-talks-in-his-personal-capacity-on-the-killing-of-cecil-a-wildlife-icon.

83   "Hunters investigate killing of Zim's best-known lion." *News24* Correspondent, July 14, 2015. www.news24.com/Green/NewsHunters-investigate-killing-of-Zims-best-known-lion-20150714.

84   Joint press statement by Zimbabwe Parks and Wildlife Management Authority and Safari Operators Association of Zimbabwe on the illegal hunt of a collared lion at Antoinette farm, Hwange District on 1 July 2015 in Gwaai Conservancy by Bushman Safaris Professional Hunter, Theo Bronkhorst. E. Chidziya (Director ZPWMA) and E. Fundira (Chairman SOAZ). Undated statement.

85   "More info about the hunt of Cecil the lion." Cruise, A. June 21, 2015. *Africa Geographic*. http://africageographic.com/blog/more-info-about-the-hunt-of-cecil-the-lion/#sthash.gJEDzVMa.dpuf.

86   An article presenting an analysis of the media activity surrounding this incident is analyzed in: "Cecil: A Moment or a Movement? Analysis of Media Coverage of the Death of a Lion, *Panthera leo*." David W. Macdonald, Kim S. Jacobsen, Dawn Burnham, Paul J. Johnson and Andrew J. Loveridge. *Animals* 2016, 6, 26; doi:10.3390/ani6050026.

87   *The Telegraph*. July 28, 2015. Alexander, H. Thornycroft, P. Laing, A. "Cecil the lion's killer revealed as American dentist." www.telegraph.co.uk/news/worldnews/africaandindianocean/zimbabwe/11767119/Cecil-the-lions-killer-revealed-as-American-dentist.html.

88   "Cecil the lion: Ricky Gervais and Cara Delevingne lead outpouring of anger after trophy hunter is identified as Walter Palmer." Heather Saul. *The Independent*. July 29, 2015. www.independent.co.uk/news/people/cecil-the-lion-ricky-gervais-and-cara-delevingne-lead-outpouring-of-anger-after-trophy-hunter-is-10423147.html.

89   "Celebrities react to Cecil the lion's death." *Hello Magazine*. July 29, 2015. www.hellomagazine.com/celebrities/2015072926477/cecil-the-lion-celebrities-twitter-outrage.

90   "Walter Palmer Is Satan': Celebrities Rage Over Cecil the Lion's Killer." Joanna Plucinska. *TIME*. July 30, 2015. http://time.com/3978108/walter-palmer-cecil-the-lion-celebrities.

91 "Peta Wants to See Cecil the Lion Killer Walter Palmer 'Extradited, Charged, and Preferably Hanged.'" Lucy Sherriff. *Huffington Post*. www.huffingtonpost. co.uk/2015/07/29/peta-want-american-lion-killer-walter-palmer-hanged_n_7892770. html?1438156052.

92 "Mia Farrow Shares Address of Dentist Who Killed the Lion Cecil in Zimbabwe, Internet Reacts." E-News. Corinne Heller. July 30, 2015. www.eonline.com/news/681607/ mia-farrow-shares-address-of-dentist-who-killed-the-lion-cecil-in-zimbabwe-internet-reacts.

93 *Jimmy Kimmel Live!*. July 28, 2015. www.youtube.com/watch?v=saHGvxFAhE0.

94 "Cecil: A Moment or a Movement? Analysis of Media Coverage of the Death of a Lion, Panthera leo." Macdonald, D.W., Jacobsen, K. Burnham, D., Johnson, P., Loveridge, A.J. Animals 2016, 6, 26; doi:10.3390/ani6050026.

95 Interview with Evan Davis, BBC *Newsnight*. July 29, 2015. http://youtu.be/ nBRaGGwqbG0

96 Measures taken by the government of Zimbabwe to improve the administration of hunting in the country. August 2, 2015. Statement issued by ZPWMA.

97 *ABC News*. August 13, 2015. Sheheri, T., Berman, T. Valiente, A. "See Photos of Black Bear Illegally Hunted by Dentist Walter Palmer Who Killed Cecil the Lion." http://abcnews.go.com/US/photos-black-bear-illegally-hunted-dentist-walter-palmer/ story?id=33067963.

98 Palmer's dental assistant, Tammy Brevik, alleged that Palmer had sexually harassed her and sued him for discrimination. Palmer denied any wrongdoing and Brevik received a settlement of $127,500. When approached for comment by the *Daily Mail*, she noted, "It's amazing how big this has become—karma is a bitch." www.dailymail.co.uk/ news/article-3180304/Karma-b-h-Woman-sexually-harassed-dentist-killed-Cecil-lion-speaks-vilified-world.html#ixzz53uIvZHuh.

99 Perhaps as part of a political power play, Grace Mugabe appeared to overrule Zimbabwe Environment Minister Oppah Muchinguri-Kashiri's call for Palmer to face prosecution. "Grace Mugabe defends Lion Killer." Mugove Tafirenyika. Nehanda Radio. August 29, 2015. http://nehandaradio.com/2015/08/29/ grace-mugabe-defends-lion-killer-walter-palmer-not-to-blame.

100 "A year later, feds still investigating Twin Cities dentist who killed Cecil the lion." Paul Walsh. *Star Tribune*. June 30, 2016. www.startribune.com/a-year-later-feds-still-investigating-twin-cities-dentist-who-killed-cecil-the-lion/385082771.

101 In response to inquiries by *Lion Hearted* editor George Rush, USFWS spokesperson, Christina Meister, stated that Walter Palmer was still under investigation for possible violation of the Lacey Act, but that no further information could be provided. January 12, 2018. Lacey Act: www.fws.gov/international/laws-treaties-agreements/ us-conservation-laws/lacey-act.html.

102 "Zimbabwe court drops charges against hunter in Cecil the lion death." Amy R. Connolly. *UPI news*. November 12, 2016. www.upi.com/Zimbabwe-court-drops-charges-against-hunter-in-Cecil-the-lion-death/3511478970996.

103 "Zimbabwean Cecil the lion hunter faces new charges." Peta Thornycroft. *The Telegraph*. September 15, 2015. www.telegraph.co.uk/news/worldnews/africaandindianocean/ zimbabwe/11867254/Zimbabwean-Cecil-the-lion-hunter-faces-new-charges.html.

104    "Sir Roger Moore on Cecil the lion: 'Hunting is a coward's pastime.'" Op-ed Roger Moore. The Telegraph. July 29, 2015. www.telegraph.co.uk/news/worldnews/africaandindianocean/zimbabwe/11771713/Cecil-the-lion-Sir-Roger-Moore-says-hunting-is-a-cowards-pastime.html.

105    "Ted Nugent hits out at 'stupid' Cecil the lion protesters, saying hunting of Africa's favorite big cat was 'legal & ESSENTIAL.'" Kieran Corcoran. July 30, 2015. Daily Mail. www.dailymail.co.uk/news/article-3179463/Ted-Nugent-hits-Cecil-lion-protesters-saying-hunting-Africa-s-favorite-lion-legal-ESSENTIAL.html.

106    "Son of Cecil the lion killed by trophy hunter." Damien Carrington. The Guardian. July 20, 2017. www.theguardian.com/environment/2017/jul/20/son-of-cecil-the-lion-killed-by-trophy-hunter.

107    "Botswana cheetah Legolas killed in 'unnecessary' attack." BBC News. October 5, 2015. www.bbc.co.uk/news/world-africa-34448465http://www.bbc.co.uk/news/world-africa-34448465.

108    "You remember Cecil the lion. But will you recall Scarface, the slain grizzly?" Karin Brulliard. The Washington Post. May 4, 2016. www.washingtonpost.com/news/animalia/wp/2016/05/04/you-remember-cecil-the-lion-but-will-you-recall-scarface-the-slain-grizzly/?utm_term=.557d74b5ffd8.

109    "Conservation inequality and the charismatic cat." Macdonald, E. Burnham, D. Hinks, A. Dickman, A. Mali, Y. Macdonald, D. Global Ecology and Conservation 3 (2015): 851–866.

110    "Blue and black or white and gold, how the dress colour you see says a lot about you." Griffin, A. The Independent. February 27, 2015. www.independent.co.uk/news/science/what-colour-is-the-dress-blue-and-black-or-white-and-gold-whatever-you-see-says-a-lot-about-you-10074490.html.

111    "When Silicon Valley Took Over Journalism. The pursuit of digital readership broke the New Republic and an entire industry." Foer, F. The Atlantic. September 2017. www.theatlantic.com/magazine/archive/2017/09/when-silicon-valley-took-over-journalism/534.

112    "Cecil the lion shares message of hope to psychic's Facebook page." Breitbart News, August 2, 2015. www.breitbart.com/big-government/2015/08/02/cecil-the-lion-shares-message-of-hope-to-animal-psychics-facebook-page.

## CHAPTER 13

113    "Trump says big game trophy ban reversal is on hold to review 'conservation facts.'" Jacobo, J. November 17, 2017. ABC News. http://abcnews.go.com/Politics/trump-tweets-big-game-trophy-ban-reversal-hold/story?id=51235849.

114    "Photos of Donald Trump's adult sons hunting in Africa resurface, spark comparisons with Walter Palmer, killer of Cecil the lion." New York Daily News. Adam Edelman. July 29, 2015. www.nydailynews.com/news/politics/pics-rump-adult-sons-hunting-game-africa-resurface-article-1.2308107.

115    A survey of 3,011 registered US voters, conducted December 2 through December 3, 2017, showed that 69 percent of Americans oppose trophy hunting and 78 percent oppose imports to the US of trophies of hunted lions or elephants.

Survey undertaken by Remington Research Group. 1420 Northwest Vivion Road. Kansas City, Missouri 64118 816-746-4410. www.RemingtonResearchGroup. com. "Survey of American electorate reveals overwhelming opposition to trophy hunting." December 7, 2017. https://blog.humanesociety.org/wayne/2017/12/ survey-american-electorate-reveals-overwhelming-opposition-trophy-hunting. html?credit=blog_post_121317_idhome-page.

116 "Economic and conservation significance of the trophy hunting industry in sub-Saharan Africa." Lindsey, P., Roulet, P. Romanach, S. Biological Conservation 2006.

117 *The Penitent Butchers. The Fauna and Flora Preservation Society.* Fitter, R. and Scott, P. Collins. Fauna and Flora Preservation Society. 1978.

118 "Performance of parks in a century of change." Cumming, D. 2004. In: *Parks in Transition.* Editor, Brian Child. Earthscan. London.

119 "The Trophy Hunting of African Lions: Scale, Current Management Practices and Factors Undermining Sustainability." Lindsey P. Balme, G. Funston, P. Henschel, P, Madzikanda, H., Hunter, L. 2013. *Plos ONE.* September 2013. Volume 8. Issue 9, e73808. Around 1.2 million square kilometers of land is set aside for hunting.

120 Palmer later denied this was the amount but refused to say whether he had paid more or less than this. Full transcript: "Walter Palmer speaks about Cecil the lion controversy." Interview with *Minneapolis Star-Tribune.* www.startribune.com/ full-transcript-walter-palmer-speaks-about-controversy/325453351.

121 "The Significance of African Lions for the Financial Viability of Trophy Hunting and the Maintenance of Wild Land." P. Lindsey et, al. *PLOS One,* 2012, Volume 7, issue 1.

122 "Financial shortfalls facing African protected areas for lion conservation." P. Lindsey et al. manuscript in preparation. November 2017.

123 Yellowstone National Park's budget is even higher at $4,100 per square kilometer per year. *Against Extinction.* 2004. Adams, W. pg. 212. Earthscan UK and USA.

124 "The Significance of African Lions for the Financial Viability of Trophy Hunting and the Maintenance of Wild Land." P. Lindsey et, al. *PLOS One,* 2012, Volume 7, issue 1.

125 "Conserving large carnivores: dollars and fence." Packer, C., Loveridge, A., Canney, S., Caro, T. et al. *Ecology Letters,* (2013) doi:10.1111/ele.12091

126 Unpublished data, WildCRU Hwange Lion Project. While 18 percent of people said they benefitted from wildlife revenues, only 5 percent said they gained any direct financial benefit, with relatively modest one-off payments averaging forty dollars. These figures are similar to those compiled by conservation economist Ivan Bond, showing that, in communities that benefit from wildlife revenues, largely from sport hunting, each household received a dividend payment of twenty dollars a year. The cost to households that lose livestock to wild predators averages between $140 and $190 per year. Clearly, these losses are not offset by revenues derived from the use of wildlife. Ivan Bond. "CAMPFIRE and the incentives for Institutional Change." In *African Wildlife and Livelihoods.* P227-243. Eds: Hulme, D. and Murphree, M. James Currey Ltd. Oxford.

127 "How big game hunting is dividing southern Africa." *BBC News.* Mark Easton. September 10, 2017. www.bbc.co.uk/news/world-africa-41163520.

128    The value of a single lion to photographic tourism in Amboseli National Park was calculated at $27,000 in 1979, based on the amount of time tourists spent watching lions (30 percent of their time) and the amount paid for safaris. Taking into account inflation using Cost Price Index (www.bls.gov/data/inflation_calculator.htm), this equates to $97,500 in 2017. "Economics and conservation in third world national parks." Western, D., & Henry, W. 1979. *BioScience*, 29(7), 414–18.

129    Philosophy professor and ethicist Marc Moffett believes that the hunting of Cecil by Palmer on a bait did not constitute "fair chase" in that "shooting an animal over a bait with an advance bow . . . is not necessarily a difficult thing." "[Palmer] wasn't himself doing the whole process. He wasn't taking advantage of his skills, his knowledge of the animals and the environment, to coax this lion into a position where he could dispatch it in a careful way." "Did Walter Palmer give Cecil the lion a 'fair chase'? We asked a hunter/philosopher." *The Washington Post*. Michael E. Miller July 31, 2015. www.washingtonpost.com/news/morning-mix/wp/2015/07/31/did-walter-palmer-give-cecil-the-lion-a-fair-chase-we-asked-a-hunterphilosopher.

130    *The Penitent Butchers*, Fitter, R. and Scott, P. 1978. Fauna and Flora Preservation Society. Collins. 48pp.

131    "Fox hunting ban looks inevitable," *BBC News*. November 17, 2004. http://news.bbc.co.uk/1/hi/uk_politics/4015075.stm.

132    *African Hunting Gazette*, Oct/Nov/Dec 2017 issue, article defending the hunting of Xanda in 2017, by Zig Macintosh. *African Outfitter*, Nov/Dec 2017 issue. Article by Ron Thompson.

133    "The Hunters' Image." Ivan Carter. *African Hunter*. Issue 109, pg 6. April 2017.

134    "House Takes Stand Against Wildlife Trafficking." US Foreign affairs committee. Press Release, November 2, 2015. https://foreignaffairs.house.gov/press-release/house-takes-stand-against-wildlife-trafficking.

135    "Australia bans hunting 'trophies' from lions entering or leaving the country." *The Guardian*. March 13, 2015. www.theguardian.com/environment/2015/mar/13/australia-bans-hunting-trophies-from-lions-entering-or-leaving-the-country.

136    "France bans imports of lion hunt trophies." *The Guardian*, November 19, 2015. www.theguardian.com/environment/2015/nov/19/france-bans-imports-of-lion-hunt-trophies.

137    Forty airlines have banned carriage of hunting trophies since the outcry over the killing of Cecil. "Major Airlines Ban Hunting Trophies." Jack Martinez, August 4, 2015. *Newsweek*. www.newsweek.com/major-airlines-ban-hunting-trophies-359589.

138    "We Can Conserve Elephants Without Hunting Them." Dan Ashe. January 4, 2018. www.aza.org/from-the-desk-of-dan-ashe/posts/statement-by-dan-ashe-on-elephant-trophy-import-ban.

139    "A Roaring Trade. The legal trade in *Panthera leo* bones of African to East South East Asia." *Plos ONE* 12 (10): e0185996. https://doi.org/10.1371/journalpone0185996.

## CHAPTER 14

140    African Lion, *Panthera leo*. The IUCN Red List of Threatened Species. Version 2017-2. www.iucnredlist.org. Downloaded on November 16, 2017.

141   A recent Pan-African aerial survey suggests there are 352,271 savannah elephants left on the continent. "African elephant numbers plummet by 30 percent, landmark survey finds." Steyn, P. *National Geographic*, August 31, 2016. Recent estimates suggest there are https://news.nationalgeographic.com/2016/08/wildlife-african-elephants-population-decrease-great-elephant-census.

142   White Rhinoceros. *Ceratotherium simum*. The IUCN Red List of Threatened Species. Version 2017-2. www.iucnredlist.org. Downloaded on 16 November 2017.

143   *United Nations, Department of Economic and Social Affairs, Population Division (2017). World Population Prospects: The 2017 Revision, Key Findings and Advance Tables.* ESA/P/WP/248. https://esa.un.org/unpd/wpp/Publications.

144   "Future threats to biodiversity and pathways to their prevention." Tilman, D., Clark, M. Williams, D., Kimmel, S., Polasky, S., Packer, C. 2017. Nature, Vol. 546, pg73. June 2017.

145   For instance, wolves in the Yellowstone ecosystem are estimated to be worth between $22 and $48 million annually to wolf-related tourism. "Status and ecological effects of the World's largest carnivores." Ripple, W. et al. 2014. *Science.* January 10, 2014. Vol. 343, Issue 6167, 1241484 (2014). http://science.sciencemag.org/content/343/6167/1241484.

146   In the mid-twentieth century, environmental philosopher Aldo Leopold wrote about what are now termed trophic cascades, recognizing in particular the role that predators play in regulating herbivore populations and the consequent effects on the wider ecosystem. In his essay "Thinking Like a Mountain" in *A Sand County Almanac and Sketches Here and There* (1949, Oxford University Press) he wrote: "I now suspect that just as a deer herd lives in mortal fear of its wolves, so does a mountain live in mortal fear of its deer. And perhaps with better cause, for while a buck pulled down by wolves can be replaced in two or three years, a range pulled down by too many deer may fail of replacement in as many decades."

147   "Trophic downgrading of planet Earth." Estes, J. *et al.*, 2011. *Science.* July 15, 2011: Vol. 333, Issue 6040, pp. 301–306. http://science.sciencemag.org/content/333/6040/301/tab-pdf.

148   "Financial shortfalls facing African protected areas for lion conservation." P. Lindsey, et al. manuscript in preparation. November 2017.

149   African Parks. www.african-parks.org/index.php/about-us/our-story.

150   "A cultural conscience for conservation." *Animals.* 2017. Volume 7, pg 52. Goode, C., Burnham, D. Macdonald, D. July 2017.

# IMAGE CREDITS